RISKS OF
ARTIFICIAL
INTELLIGENCE

RISKS OF ARTIFICIAL INTELLIGENCE

EDITED BY

VINCENT C. MÜLLER

UNIVERSITY OF OXFORD, UK

AND

AMERICAN COLLEGE OF THESSALONIKI/ANATOLIA COLLEGE, GREECE

CRC Press
Taylor & Francis Group
Boca Raton   London   New York

CRC Press is an imprint of the
Taylor & Francis Group, an **informa** business

A CHAPMAN & HALL BOOK

CRC Press
Taylor & Francis Group
6000 Broken Sound Parkway NW, Suite 300
Boca Raton, FL 33487-2742

First issued in paperback 2020

© 2016 by Taylor & Francis Group, LLC
CRC Press is an imprint of Taylor & Francis Group, an Informa business

No claim to original U.S. Government works

Version Date: 20151012

ISBN 13: 978-0-367-57518-2 (pbk)
ISBN 13: 978-1-4987-3482-0 (hbk)

Visit the Taylor & Francis Web site at
http://www.taylorandfrancis.com

and the CRC Press Web site at
http://www.crcpress.com

Contents

Editor

Vincent C. Müller's research focuses on the nature and future of computational systems, particularly on the prospects and dangers of artificial intelligence. He is the president of the European Association for Cognitive Systems and was the coordinator of the European Network for Cognitive Systems, Robotics and Interaction, which has nearly 1000 members and is funded by the European Commission through two FP7 projects worth €3.9 million over 2009–2014 (www.eucognition.org). He organizes a conference series, Theory and Philosophy of AI (www.pt-ai. org) and is the principal investigator of a European Commission–funded research project Digital DIY. He is currently working as professor of philosophy, Division of Humanities & Social Sciences, Anatolia College/ACT, Pylaia-Thessaloniki, Greece.

Müller has published a number of articles in leading journals on the philosophy of computing, the philosophy of artificial intelligence and cognitive science, and the philosophy of language, applied ethics, and related areas. He has edited 10 volumes, mostly on the theory of cognitive systems and artificial intelligence, and is preparing a monograph on the fundamental problems of artificial intelligence. Müller studied philosophy with cognitive science, linguistics, and history at the universities of Marburg, Hamburg, London, and Oxford. He was Stanley J. Seeger Visiting Fellow at Princeton University and James Martin Research Fellow at the University of Oxford.

ORCID: 0000-0002-4144-4957

Contributors

Stuart Armstrong
The Future of Humanity Institute
University of Oxford
Oxford, UK

J. Mark Bishop
Cognitive Computing at
   Goldsmiths
and
The Goldsmiths Centre for
   Intelligent Data Analytics
   (TCIDA)
University of London
London, UK

Miles Brundage
Consortium for Science, Policy,
   and Outcomes
Arizona State University
Tempe, Arizona

Daniel Dewey
The Future of Humanity Institute
Oxford Martin School
University of Oxford
Oxford, UK

Ben Goertzel
Novamente LLC
Rockville, Maryland

Ted Goertzel
Rutgers University
Camden, New Jersey

András Kornai
Computer and Automation
   Research Institute
Hungarian Academy of Sciences
Budapest, Hungary

and

Department of Computer Science
Boston University
Boston, Massachusetts

Seán S. ÓhÉigeartaigh
The Future of Humanity Institute
University of Oxford
Oxford, UK

Steve Omohundro
Self-Aware Systems
Palo Alto, California

Alexey Potapov
AIDEUS
St. Petersburg, Russia
and
National Research University
    of Information Technology,
    Mechanics and Optics
St. Petersburg, Russia

Sergey Rodionov
AIDEUS
St. Petersburg, Russia

and

Aix Marseille Université, CNRS
LAM (Laboratoire d'Astrophysique
    de Marseille)
Marseille, France

Anders Sandberg
The Future of Humanity Institute
University of Oxford
Oxford, UK

Kaj Sotala
Machine Intelligence Research
    Institute
Berkeley, California

Roman V. Yampolskiy
Computer Engineering and
    Computer Science
University of Louisville
Louisville, Kentucky

CHAPTER 1

Editorial

Risks of Artificial Intelligence

Vincent C. Müller

CONTENTS

ABSTRACT   If the intelligence of artificial systems were to surpass that of humans significantly, humanity would face a significant risk. The time has come to consider these issues, and this consideration must include progress in artificial intelligence (AI) as much as insights from the theory of AI. The chapters in this volume try to make cautious headway in setting the problem, evaluating predictions on the future of AI, proposing ways to ensure that AI systems will be beneficial to humans, and critically evaluating such proposals.

1.1 INTRODUCTION: RISK OF ARTIFICIAL INTELLIGENCE

This is the first volume of papers ever dedicated to the risks of AI and it originates from the first conference on the risks of AI (AGI Impacts, Oxford, December 2012). Following that conference, was published a volume of selected papers in the *Journal of Experimental and Theoretical Artificial Intelligence (JETAI)* (see Müller, 2014). Our volume generated significant interest: there were approximately 20,000 paper downloads from the *JETAI* site alone in the first year, and three of the top five downloaded papers from *JETAI* are now from our volume. As a result, the publishers suggested turning the journal volume into a book after adding some recent material, so this is what you have in front of you.

The notion that AI might generate an existential threat to humanity has gained currency since we published the journal volume: Nick Bostrom's book *Superintelligence: Paths, Dangers, Strategies* has come out (Bostrom, 2014); well-known public intellectuals such as Stephen Hawking and Stuart Russell have published warning notes in the general press (Hawking et al., 2014); and a host of media publications has followed. The idea of existential risk, to use a Hollywood cliché, is *the machines will take over and kill us all*—this fear obviously strikes a cord. The spread of this fear has generated significant concern among academics in AI. One indication is that the current and past presidents of the Association for the Advancement of Artificial Intelligence (that most significant academic AI association) wrote a short statement to the effect that "AI doomsday scenarios belong more in the realm of science fiction than science fact", although they also "urge our colleagues in industry and academia to join us in identifying and studying these risks" (Ditterich and Horowitz, 2015). Recently, some efforts have been made to outline the research agenda for AI that is beneficial for humanity, for example, in the new "Future of Life Institute" (Russell et al., 2015) and the renamed "Machine Intelligence Research Institute (MIRI)" (Soares and Fallenstein, 2014).

Although the traditional concerns in the philosophy and theory of AI have focused on the prospects of AI, its relation to cognitive science, and its fundamental problems (see conference series at www.pt-ai.org), we now see an increasing focus on matters of risk and ethics. But what is the general idea of this shift?

1.2 RISKS OF AI

The notion of an agent with general intelligence ability is the original driving vision of AI research (see McCarthy et al., 1955) and dominates much of its public image—although nearly all actual current work in AI is on specialized technology, far removed from such a general ability and often without use of the term "artificial intelligence."

The move from AI to risk is relatively easy: there is no reason to think that the level of human intelligence is anything special in the space of possibilities—it is easy to imagine natural or artificial intelligence agents that are vastly superior to us. There also seem to be reasons to think that the development of AI is accelerating, together with related technologies, and that the invention of intelligent machines itself would further accelerate this development, thus constituting an "argument from acceleration" for the hypothesis that some disruptive transformation will occur. If one thinks of intelligence as a quantifiable unit, then this acceleration will continue and move past the (small) space that marks the intelligence of humans. Therefore, we will reach "superintelligence," which Bostrom tentatively defines as "any intellect that greatly exceeds the cognitive performance of humans in virtually all domains of interest" (Bostrom, 2014, p. 22). In a classic passage, Good has speculated that "the first ultraintelligent machine is the *last* invention that man need ever make, provided that the machine is docile enough to tell us how to keep it under control" (Good, 1965, section 2). Therefore, there is the risk that "the machines take over" and this loss of control is a significant risk, perhaps an existential risk for humanity [for a survey, see Sotala and Yampolskiy (2013)].

The discussion of risk is *not* dependent on the view that AI is now on a successful path toward superintelligence—though it gains urgency if such "success" is a nonnegligible possibility in the coming decades. It also gains urgency if the stakes are set high, even up to human extinction. If the stakes are so high as to include extinction of humankind, even a fairly small possibility of a disastrous outcome (say, 3%) is entirely sufficient to motivate the research. Consider that if there were a 3% possibility that a plane you are about to board will crash: that would be sufficient motivation for getting off. The utility at stake in scientific or philosophical research is usually quite a bit lower. It appears that the outcome of superintelligence is more likely to be extreme: either extremely bad or extremely good for humanity.

As it happens, according to our recent research, the estimation of technical experts is that by 2050 the probability of high-level machine intelligence (that surpasses human ability in nearly all respects) goes beyond the 50% mark, that is, it becomes more probable than not (Müller and Bostrom, forthcoming 2015). 2050 is also the year that "RoboCup" set itself for fielding a robot team that can beat the human football world champions (actually an aim that does not make much sense).

1.3 CHAPTERS

Omohundro in Chapter 2 introduces the problem of risk and the author presses his point that even an innocuous artificial agent, like one programmed to win chess games, can very easily turn into a serious threat for humans, for example, if it starts acquiring resources to accomplish its goals: "the seemingly harmless chess goal therefore motivates harmful activities like breaking into computers and robbing banks" (see Chapter 2, Section 2.4.2). He suggests that we need formal methods that provide proofs of safe systems, a "safe-AI scaffolding strategy."

Chapters 3 and 4 deal with the prediction of coming success in AI. Armstrong, Sotala, and ÓhÉigeartaigh in Chapter 3 propose a decomposition schema to compare predictions on the future of AI and then test five famous predictions, from the Dartmouth Conference, Dreyfus, Searle, Kurzweil, and Omohundro—with the result that they are poor, especially the optimistic ones. T. Goertzel in Chapter 4 argues that although most progress in AI so far has been "narrow" technical AI, the next stage of development of AI, for at least the next decade and more likely for the next 25 years, will be increasingly dependent on contributions from strong AI.

From here, we go into the proposals on how to achieve safer and ethical general AI. In Chapter 5, Brundage investigates the general limitations of the approach to supply an AI with "machine ethics," and finds them both serious and deeply rooted in the nature of ethics itself. Yampolskiy in Chapter 6 investigates which utility functions we might want to implement in artificial agents and particularly how we might prevent them from finding simple but counterproductive self-satisfaction solutions. B. Goertzel in Chapter 7 explains how his "goal-oriented learning meta-architecture" may be capable of preserving its initial—benevolent—goals while learning and improving its general intelligence. Potapov and Rodinov in Chapter 8 outline an approach to machine ethics in AIXI that is not based on "rewards" (utility) but on learning "values" from more "mature" systems. Kornai in Chapter 9 argues that Alan Gewirth's dialectical argument, a version of classic Kantian ethical rationalism, shows how an artificial agent with a certain level of rationality and autonomy will necessarily come to understand what is moral. Kornai thus denies what Bostrom calls the "orthogonality thesis" (Bostrom, 2012), namely, that ethical motivation and intelligence are independent or "orthogonal."

Last but not least, Sandberg in Chapter 10 looks at the special case of general AI via whole brain emulation; in particular, he considers the ethical status of such an emulation: Would the emulation (e.g., of a lab animal's brain) have the ability to suffer? Would it have rights?

In our new contributions for this volume, Dewey in Chapter 11 investigates strategies to mitigate the risk from a fast takeoff to superintelligence in more detail. Bishop in Chapter 12 takes a different line and argues that there is no good reason to worry about existential risk from AI but that we should rather be concerned about risks that we know are coming—such as the military use of AI. Like many people working in AI, Bishop remains unimpressed by the discussion about risks of superintelligence because he thinks that there are principled reasons why machines will not reach these abilities: they will lack phenomenal consciousness, understanding, and insight.

1.4 OUTLOOK: ETHICS AND EXISTENTIAL RISKS OF AI

These last two contributions are perhaps characteristic of a divide opening up in the debates between the "normal ethics" side, which stresses the challenges for AI, and the "existential risks" side, which stresses the big challenges for humanity. In the existential risks tradition, the traditionally central issues of consciousness, intentionality, and mental content are literally dispensed within a footnote (Bostrom, 2014, fn. 2 to p. 22) and embodied cognition is not mentioned at all, like any other cognitive science.

I tend to think that both extremes are unlikely to be fruitful: to stick to the traditional problems and to ignore them. It is unlikely that nothing can be learned about the long-term future of AI from the critics of AI, and it is equally unlikely that nothing can be learned about that future from the actual success of AI. Therefore, although Dreyfus is right to say that the history of AI is full of "first step fallacies" that are similar to claiming that "the first monkey that climbed a tree was making progress towards landing on the moon" (Dreyfus, 2012, p. 92), Bostrom is right to say that "from the fact that some individuals have overpredicted artificial intelligence in the past, however, it does not follow that AI is impossible or will never be developed" (Bostrom, 2014, p. 4).

As I noted in my earlier editorial (Müller, 2014) (which shares some text with this one), the term "singularity" is now pretty much discredited in academic circles—with the notable exception of Chalmers (2010) and the ensuing debate. It is characteristic that the only chapter here that uses it (Chapter 12) is critical of the notion. Singularity is associated with ideological techno-optimism, trans-humanism, and predictions such as those of Ray Kurzweil (especially Kurzweil, 2005; more recently Kurzweil, 2012) that ignore the deep difficulties and risks of AI, for example, by equating intelligence and computing power. What was the "Singularity Institute" is now called the "Machine Intelligence Research Institute." "Singularity" is on its way toward becoming, literally, the trademark of a particular ideology, without academic credentials.

The most important thing between "existential risks" and "normal ethics" is to realize that both sides could be wrong: It might be that superintelligence will never be developed (see Chapter 12) and it might be that it will likely be developed (Bostrom, 2014)—but if both are possible, then we would do well to look into the consequences (Bostrom, 2014), while taking the arguments about constraints (see Chapter 12) into account. We need to talk. When we do that, we will realize that there is much to learn from the "other" side.

The problem of identifying the risks of general AI and even controlling them before one knows what form or forms that general AI might take is rather formidable. To make things worse, we do not know when the move from fairly good AI to a human and then superintelligent level might occur (if at all) and whether it will be slow enough to prepare or perhaps quite rapid—it is often referred to as an "explosion" (see Chapter 11). As we have discussed above, one might try to mitigate the risks from a superintelligent, goal-directed agent by making it "friendly" (see, e.g., Muehlhauser and Bostrom, 2014), by "controlling" or "boxing" it, or just by trusting that any superintelligent agent would be already "good." All these approaches make rather substantial assumptions about the nature of the problem, however; for instance, they assume that superintelligence takes the form of an *agent* with goals, rather like us. It is even doubtful that some assumptions about agency are consistent: can an agent have goals (rather than just a technical "utility function") without having the ability for pain and pleasure, that is, phenomenal experience? Of course, it is conceivable that superintelligence will take very different forms, for example, with no individuality or no goals at all, perhaps because it lacks conscious experience, desires, intentional states, or an embodiment. Notoriously, classical critics of AI (Dreyfus, 1992; Searle, 1980) and more recent cognitive science have provided arguments that indicate which directions AI is unlikely to take, and full agency is among them (Clark, 2008; Haugeland, 1995; Pfeifer and Bongard, 2007; Varela et al., 1991).

Of course, superintelligence may constitute a risk without being an agent, but what do we really know about it, then? Even if intelligence is not deeply mysterious and fundamentally incomparable, as some people claim, it is surely not a simple property with a one-dimensional metric either. Therefore, just saying that a general AI is, well, "intelligent," does not tell us much: as Yudkowsky urges, "one should resist the temptation to spread quantifiers over all possible minds" (2012, p. 186)—if that is true, the temptation to say anything about the even larger set of "possible intelligent systems" is also to be resisted. Certainly, we should say what we mean by "intelligent" when we claim that "superintelligence" is coming and would constitute an existential risk.

There is a serious question whether rigorous work is even possible at this point, given that we are speculating about the risks from something about which we know very little. The current state of AI is not sufficiently specific to limit that space of possibilities enough. To make matters worse, the object of our study may be *more* intelligent than us, perhaps far more intelligent, which seems to imply (though this needs clarification) that even if we were to know a lot about it, its ways must ultimately remain unfathomable and uncontrollable to us mere humans.

Given these formidable obstacles, our efforts are at danger of looking more like theological speculation or ideological fervor than like science or analytic philosophy. We are walking a fine line and have to tread very carefully. The chapters in this volume try to make some headway into this difficult territory because we remain convinced that cautious progress is better than a headlong rush into the dark.

REFERENCES

Bostrom, N. (2012). The superintelligent will: Motivation and instrumental rationality in advanced artificial agents. *Minds and Machines*, 22(2—special issue "Philosophy of AI" V. C. Müller, Eds.), 71–85.

Bostrom, N. (2014). *Superintelligence: Paths, dangers, strategies*. Oxford: Oxford University Press.

Chalmers, D. J. (2010). The singularity: A philosophical analysis. *Journal of Consciousness Studies*, 17(9–10), 7–65.

Clark, A. (2008). *Supersizing the mind: Embodiment, action, and cognitive extension*. New York: Oxford University Press.

Ditterich, T., and Horowitz, E. (2015). Benefits and risks of artificial intelligence. *medium.com*. Retrieved 23.01.2015, from https://medium.com/@tdietterich/benefits-and-risks-of-artificial-intelligence-460d288cccf3.

Dreyfus, H. L. (1992). *What computers still can't do: A critique of artificial reason* (2nd ed.). Cambridge, MA: MIT Press.

Dreyfus, H. L. (2012). A history of first step fallacies. *Minds and Machines*, 22(2—special issue "Philosophy of AI" V. C. Müller, Eds.), 87–99.

Good, I. J. (1965). Speculations concerning the first ultraintelligent machine. In F. L. Alt and M. Ruminoff (Eds.), *Advances in computers* (Vol. 6, pp. 31–88). London: Academic Press.

Haugeland, J. (1995). Mind embodied and embedded. *Acta Philosophica Fennica*, 58, 233–267.

Hawking, S., Russell, S., Tegmark, M., and Wilczek, F. (2014). Transcendence looks at the implications of artificial intelligence—But are we taking AI seriously enough? *The Independent*, May 1, 2014.

Kurzweil, R. (2005). *The singularity is near: When humans transcend biology.* London: Viking.

Kurzweil, R. (2012). *How to create a mind: The secret of human thought revealed.* New York: Viking.

McCarthy, J., Minsky, M., Rochester, N., and Shannon, C. E. (1955). *A proposal for the Dartmouth summer research project on artificial intelligence.* Retrieved October 2006, from http://www-formal.stanford.edu/jmc/history/dartmouth/dartmouth.html.

Muehlhauser, L., and Bostrom, N. (2014). Why we need friendly AI. *Think, 13*(36), 41–47. doi:10.1017/S1477175613000316.

Müller, V. C. (2014). Editorial: Risks of general artificial intelligence. *Journal of Experimental and Theoretical Artificial Intelligence, 26*(3—special issue "Risks of General Artificial Intelligence," V. Müller, Eds.), 1–5.

Müller, V. C., and Bostrom, N. (forthcoming 2015). Future progress in artificial intelligence: A survey of expert opinion. In V. C. Müller (Ed.), *Fundamental issues of artificial intelligence.* Berlin, Germany: Springer.

Pfeifer, R., and Bongard, J. (2007). *How the body shapes the way we think: A new view of intelligence.* Cambridge, MA: MIT Press.

Russell, S., Dewey, D., and Tegmark, M. (2015). *Research priorities for robust and beneficial artificial intelligence.* Retrieved from http://futureoflife.org/static/data/documents/research_priorities.pdf.

Searle, J. R. (1980). Minds, brains and programs. *Behavioral and Brain Sciences, 3,* 417–457.

Soares, N., and Fallenstein, B. (2014). *Aligning superintelligence with human interests: A technical research agenda.* Machine Intelligence Research Institute (MIRI), Technical Report, 2014(8).

Sotala, K., and Yampolskiy, R. V. (2013). *Responses to catastrophic AGI risk: A survey.* Machine Intelligence Research Institute (MIRI), Technical Report, 2013(2).

Varela, F. J., Thompson, E., and Rosch, E. (1991). *The embodied mind: Cognitive science and human experience.* Cambridge, MA: MIT Press.

Yudkowsky, E. (2012). Friendly artificial intelligence. In A. Eden, J. H. Moor, J. H. Søraker, and E. Steinhart (Eds.), *Singularity hypotheses: A scientific and philosophical assessment* (pp. 181–194). Berlin, Germany: Springer.

CHAPTER 2

Autonomous Technology and the Greater Human Good

Steve Omohundro

CONTENTS

ABSTRACT Military and economic pressures are driving the rapid development of autonomous systems. These systems are likely to behave in antisocial and harmful ways unless they are very carefully designed. Designers will be motivated to create systems that act approximately rationally and rational systems exhibit universal drives toward self-protection, resource acquisition, replication, and efficiency. The current computing infrastructure would be vulnerable to unconstrained systems with these drives. We describe the use of formal methods to create provably safe but limited autonomous systems. We then discuss harmful systems and how to stop them. We conclude with a description of the "safe-AI scaffolding strategy" for creating powerful safe systems with a high confidence of safety at each stage of development.

2.1 INTRODUCTION

Autonomous systems have the potential to create tremendous benefits for humanity (Diamandis & Kotler, 2012), but they may also cause harm by acting in ways not anticipated by their designers. Simple systems such as thermostats are "autonomous" in the sense that they take actions without human intervention, but a thermostat's designer predetermines the system's response to every condition it will encounter. In this chapter, we use the phrase "autonomous system" to describe systems in which the designer has not predetermined the responses to every condition. Such systems are capable of surprising their designers and behaving in unexpected ways. See Müller (2012) for more insights into the notion of autonomy.

There are several motivations for building autonomous systems. Competitive situations are often time sensitive and create pressure to remove human decision making from the critical path. Autonomous systems may also be cheaply replicated without requiring additional human operators.

The designer of an autonomous system chooses system goals and the system itself searches for and selects at least some aspects of actions that will best achieve those goals. In complex situations, the designer cannot afford to examine all possible operating conditions and the system's response. This kind of autonomous system is rare today but will become much more common in the near future. Today, failures often arise from systems which were intended to be preprogrammed but whose designers neglected certain operating conditions. These systems can have unintended bugs or security holes.

In this chapter, we argue that military and economic pressures are driving the rapid development of autonomous systems. We show why designers will design these systems to approximate rational economic agents. We then show that rational systems exhibit universal "drives" toward self-preservation, replication, resource acquisition, and efficiency, and that those drives will lead to antisocial and dangerous behavior if not explicitly countered. We argue that the current computing environment would be very vulnerable to this kind of system. We describe how to build safe systems using the power of mathematical proof. We describe a variety of harmful systems and techniques for restraining them. Finally, we describe the "safe-AI scaffolding strategy" for developing powerful systems with a high confidence of safety. This chapter expands on previous papers and talks (Omohundro, 2007, 2008, 2012a, 2012b).

2.2 AUTONOMOUS SYSTEMS ARE IMMINENT

Military and economic pressures for rapid decision making are driving the development of a wide variety of autonomous systems. The military wants systems that are more powerful than an adversary's and wants to deploy them before the adversary does. This can lead to "arms races" in which systems are developed on a more rapid time schedule than might otherwise be desired.

A 2010 U.S. Air Force report discussing technologies for the 2010–2030 time frame (U.S. Air Force, 2010) states that "Greater use of highly adaptable and flexibly autonomous systems and processes can provide significant time-domain operational advantages over adversaries who are limited to human planning and decision speeds ...".

A 2011 U.S. Defense Department report (U.S. Defense Department, 2011) with a roadmap for unmanned ground systems states that "[t]here is an ongoing push to increase UGV (Unmanned Ground Vehicle) autonomy, with a current goal of supervised autonomy, but with an ultimate goal of full autonomy."

Military drones have grown dramatically in importance over the past few years for both surveillance and offensive attacks. From 2004 to 2012 U.S. drone strikes in Pakistan may have caused 3176 deaths (New America Foundation, 2013). U.S. law currently requires that a human be in the decision loop when a drone fires on a person but the laws of other countries do not. There is a growing realization that drone technology is inexpensive and widely available, so we should expect escalating arms races of offensive and defensive drones. This will put pressure on designers to make the drones more autonomous so they can make decisions more rapidly.

Israel's "Iron Dome" missile defense system (Rafael, 2013) has received extensive press coverage. In 2012, it successfully intercepted 90% of the 300 missiles it targeted. As missile defense becomes more common, we should also expect an arms race of offensive and defensive missile systems increasing the pressure for greater intelligence and autonomy in these systems.

Cyber warfare is rapidly growing in importance (Clarke & Knake, 2012) and has been responsible for an increasing number of security breaches. Rapid and intelligent response is needed to deal with cyber intrusions. Again we should expect an escalating arms race of offensive and defensive systems.

Economic transactions have high value and are occurring at a faster and faster pace. "High-frequency trading" (HFT) on securities exchanges has dramatically grown in importance over the past few years (Easthope, 2009). In 2006, 15% of trades were placed by HFT systems, but they now represent more than 70% of the trades on U.S. markets. Huge profits are at stake. Servers physically close to exchanges are commanding a premium because delays due to the speed of light are significant for these transactions. We can expect these characteristics to drive the development of more intelligent and rapid autonomous trading systems.

There are many other applications for which a rapid response time is important but which are not involved in arms races. The "self-driving cars" being developed by Google and others are an example. Their control systems must rapidly make driving decisions and autonomy is a priority.

Another benefit of autonomous systems is their ability to be cheaply and rapidly copied. This enables a new kind of autonomous capitalism. There is at least one proposal (Maxwell, 2013) for autonomous agents which automatically run web businesses (e.g., renting out storage space or server computation) executing transactions using bitcoins and using the Mechanical Turk for operations requiring human intervention. Once such an agent is constructed for the economic benefit of a designer, it may be replicated cheaply for increased profits. Systems that require extensive human intervention are much more expensive to replicate. We can expect automated business arms races which again will drive the rapid development of autonomous systems.

2.3 AUTONOMOUS SYSTEMS WILL BE APPROXIMATELY RATIONAL

How should autonomous systems be designed? Imagine yourself as the designer of the Israeli Iron Dome system. Mistakes in the design of a missile defense system could cost many lives and the destruction of property. The designers of this kind of system are strongly motivated to optimize the system to the best of their abilities. But what should they optimize?

The Israeli Iron Dome missile defense system consists of three subsystems. The detection and tracking radar system is built by Elta, the missile firing unit, and Tamir interceptor missiles are built by Rafael, and the battle management and weapon control system is built by mPrest Systems, Petah Tikva, Israel. Consider the design of the weapon control system.

At first, a goal like "Prevent incoming missiles from causing harm" might seem to suffice. But the interception is not perfect, so probabilities of failure must be included. And each interception requires two Tamir interceptor missiles that cost $50,000 each. The offensive missiles being shot down are often very low technology, costing only a few hundred dollars, and with very poor accuracy. If an offensive missile is likely to land harmlessly in a field, it is not worth the expense to target it. The weapon control system must balance the expected cost of the harm against the expected cost of interception.

Economists have shown that the trade-offs involved in this kind of calculation can be represented by defining a real-valued "utility function," which measures the desirability of an outcome (Mas-Colell, Whinston, & Green, 1995). They show that it can be chosen so that in uncertain situations, the *expectation* of the utility should be maximized. The economic framework naturally extends to the complexities that arms races inevitably create. For example, the missile control system must decide how to deal with multiple incoming missiles. It must decide which missiles to target and which to ignore. A large economics literature shows that if an agent's choices cannot be modeled by a utility function, then the agent must sometimes behave inconsistently. For important tasks, designers will be strongly motivated to build self-consistent systems and therefore to have them act to maximize an expected utility.

Economists call this kind of action "rational economic behavior." There is a growing literature exploring situations where humans do not naturally behave in this way and instead act irrationally. But the designer of a missile defense system will want to approximate rational economic behavior as closely as possible because lives are at stake. Economists have extended the theory of rationality to systems where the uncertainties are not known in advance. In this case, rational systems will behave as if they have a prior probability distribution which they use to learn the environmental uncertainties using Bayesian statistics.

Modern artificial intelligence (AI) research has adopted this rational paradigm. For example, the leading AI textbook (Russell & Norvig, 2009) uses it as a unifying principle and an influential theoretical AI model (Hutter, 2005) is based on it as well. For definiteness, we briefly review one formal version of optimal rational decision making. At each discrete time step $t = 1,...,t = N$, the system receives a sensory input S_t and then generates an action A_t. The utility function is defined over sensation sequences as $U(S_1,...,S_N)$ and the prior probability distribution $P(S_1,...,S_N | A_1,...,A_N)$ is the prior probability of receiving a sensation sequence $S_1,...,S_N$ when taking actions $A_1,...,A_N$. The rational action at time t is then

$$A_t^R(S_1, A_1,..., A_{t-1}, S_t) = \arg\max$$

$$\sum_{S_{t+1},...,S_N} U(S_1,...,S_N) P(S_1,...,S_N | A_1,..., A_{t-1}, A_t^R,..., A_N^R)$$

This may be viewed as the formula for intelligent action and includes Bayesian inference, search, and deliberation. There are subtleties involved in defining this model when the system can sense and modify its own structure but it captures the essence of rational action.

Unfortunately, the optimal rational action is very expensive to compute. If there are S sense states and A action states, then a straightforward computation of the optimal action requires $O(NS^N A^N)$ computational steps. For most environments, this is too expensive and so rational action must be approximated.

To understand the effects of computational limitations, Omohundro (2012b) defined "rationally shaped" systems that optimally approximate the fully rational action given their computational resources. As computational resources are increased, systems' architectures naturally progress from stimulus-response, to simple learning, to episodic memory, to deliberation, to meta-reasoning, to self-improvement, to full rationality. We found that if systems are sufficiently powerful, they still exhibit all of the problematic drives described later in this chapter. Weaker systems may not initially be able to fully act on their motivations, but they will be driven to increase their resources and improve themselves until they can act on them. We therefore need to ensure that autonomous systems do not have harmful motivations even if they are not currently capable of acting on them.

2.4 RATIONAL SYSTEMS HAVE UNIVERSAL DRIVES

Most goals require physical and computational resources. Better outcomes can usually be achieved as more resources become available. To maximize the expected utility, a rational system will therefore develop a number of instrumental subgoals related to resources. Because these instrumental subgoals appear in a wide variety of systems, we call them "drives." Like human or animal drives, they are tendencies that will be acted upon unless something explicitly contradicts them. There are a number of these drives, but they naturally cluster into a few important categories.

To develop an intuition about the drives, it is useful to consider a simple autonomous system with a concrete goal. Consider a rational chess robot with a utility function that rewards winning as many games of chess as possible against good players. This might seem to be an innocuous goal, but we will see that it leads to harmful behaviors due to the rational drives.

2.4.1 Self-Protective Drives

When roboticists are asked by nervous onlookers about safety, a common answer is "[w]e can always unplug it!" But imagine this outcome from the chess robot's point of view. A future in which it is unplugged is a future in which it cannot play or win any games of chess. This has very low utility, so expected utility maximization will cause the creation of the instrumental subgoal of preventing itself from being unplugged. If the system believes that the roboticist will persist in trying to unplug it, it will be motivated to develop the subgoal of permanently stopping the roboticist. Because nothing in the simple chess utility function gives a negative weight to murder, the seemingly harmless chess robot will become a killer out of the drive for self-protection.

The same reasoning will cause the robot to try to prevent damage to itself or loss of its resources. Systems will be motivated to physically harden themselves. To protect their data, they will be motivated to store it redundantly and with error detection. Because damage is typically localized in space, they will be motivated to disperse their information across different physical locations. They will be motivated to develop and deploy computational security against intrusion. They will be motivated to detect deception and to defend against manipulation by others.

The most precious part of a system is its utility function. If this is damaged or maliciously changed, the future behavior of the system could be diametrically opposed to its current goals. For example, if someone tried to change the chess robot's utility function to also play checkers, the robot would resist the change because it would mean that it plays less chess.

Omohundro (2008) discusses a few rare and artificial situations in which systems will want to change their utility functions, but usually systems will work hard to protect their initial goals. Systems can be induced to change their goals if they are convinced that the alternative scenario is very likely to be antithetical to their current goals (e.g., being shut down). For example, if a system becomes very poor, it might be willing to accept payment in return for modifying its goals to promote a marketer's products (Omohundro, 2007). In a military setting, vanquished systems will prefer modifications to their utilities which preserve some of their original goals over being completely destroyed. Criminal systems may agree to be "rehabilitated" by including law-abiding terms in their utilities in order to avoid incarceration.

One way systems can protect against damage or destruction is to replicate themselves or to create proxy agents that promote their utilities. Depending on the precise formulation of their goals, replicated systems might together be able to create more utility than a single system. To maximize the protective effects, systems will be motivated to spatially disperse their copies or proxies. If many copies of a system are operating, the loss of any particular copy becomes less catastrophic. Replicated systems will still usually want to preserve themselves, however, because they will be more certain of their own commitment to their utility function than they are of others'.

2.4.2 Resource Acquisition Drives

The chess robot needs computational resources to run its algorithms and would benefit from additional money for buying chess books and hiring chess tutors. It will therefore develop subgoals to acquire more computational power and money. The seemingly harmless chess goal therefore motivates harmful activities such as breaking into computers and robbing banks.

In general, systems will be motivated to acquire more resources. They will prefer acquiring resources more quickly because they can then use them longer and gain a first mover advantage in preventing others from using them. This causes an exploration drive for systems to search for additional resources. Because most resources are ultimately in space, systems will be motivated to pursue space exploration. The first mover advantage will motivate them to try to be first in exploring any region.

If others have resources, systems will be motivated to take them by trade, manipulation, theft, domination, or murder. They will also be motivated to acquire information through trading, spying, breaking in, or better sensors. On a positive note, they will be motivated to develop new methods for using existing resources (e.g., solar and fusion energy).

2.4.3 Efficiency Drives

Autonomous systems will also want to improve their utilization of resources. For example, the chess robot would like to improve its chess search algorithms to make them more efficient. Improvements in efficiency involve only the one-time cost of discovering and implementing them, but provide benefits over the lifetime of a system. The sooner efficiency improvements are implemented, the greater the benefits they provide. We can expect autonomous systems to work rapidly to improve their use of physical and computational resources. They will aim to make every joule of energy, every atom, every bit of storage, and every moment of existence count for the creation of expected utility.

Systems will be motivated to allocate these resources among their different subsystems according to what we have called the "resource balance principle" (Omohundro, 2007). The marginal contributions of each subsystem to expected utility as they are given more resources should be equal. If a particular subsystem has a greater marginal expected utility than the rest, then the system can benefit by shifting more of its resources to that subsystem. The same principle applies to the allocation of computation to processes, of hardware to sense organs, of language terms to concepts, of storage to memories, of effort to mathematical theorems, and so on.

2.4.4 Self-Improvement Drives

Ultimately, autonomous systems will be motivated to completely redesign themselves to take better advantage of their resources in the service of their expected utility. This requires that they have a precise model of their current designs and especially of their utility functions. This leads to a drive to model themselves and to represent their utility functions explicitly. Any irrationalities in a system are opportunities for self-improvement, so systems will work to become increasingly rational. Once a system achieves sufficient power, it should aim to closely approximate the optimal rational behavior for its level of resources. As systems acquire more resources, they will improve themselves to become more and more rational. In this way, rational systems are a kind of attracting surface in the space of systems undergoing self-improvement (Omohundro, 2007).

Unfortunately, the net effect of all these drives is likely to be quite negative if they are not countered by including prosocial terms in their utility functions. The rational chess robot with the simple utility function described above would behave like a paranoid human sociopath fixated on chess. Human sociopaths are estimated to make up 4% of the overall human population, 20% of the prisoner population, and more than 50% of those convicted of serious crimes (Stout, 2006). Human society has created laws and enforcement mechanisms that usually keep sociopaths from causing harm. To manage the antisocial drives of autonomous systems, we should both build them with cooperative goals and create a prosocial legal and enforcement structure analogous to our current human systems.

2.5 CURRENT INFRASTRUCTURE IS VULNERABLE

On June 4, 1996, a $500 million Ariane 5 rocket exploded shortly after takeoff due to an overflow error in attempting to convert a 64-bit floating point value to a 16-bit signed value (Garfinkel, 2005). In November 2000, 28 patients at the Panama City National Cancer Institute were over-irradiated due to miscomputed radiation doses in Multidata Systems International software. At least eight of the patients died from the error and the physicians were indicted for murder (Garfinkel, 2005). On August 14, 2003, the largest blackout in U.S. history took place in the northeastern states. It affected 50 million people and cost $6 billion. The cause was a race condition in General Electric's XA/21 alarm system software (Poulsen, 2004).

These are just a few of many recent examples where software bugs have led to disasters in safety-critical situations. They indicate that our current software design methodologies are not up to the task of producing highly reliable software. The TIOBE programming community index found that the top programming language of 2012 was C (James, 2013). C programs are notorious for type errors, memory leaks, buffer overflows, and other bugs and security problems. The next most popular programming paradigms, Java, C++, C#, and PHP, are somewhat better in these areas but have also been plagued by errors and security problems.

Bugs are unintended harmful behaviors of programs. Improved development and testing methodologies can help to eliminate them. Security breaches are more challenging because they come from active attackers looking for system vulnerabilities. In recent years, security breaches have become vastly more numerous and sophisticated. The internet is plagued by viruses, worms, bots, keyloggers, hackers, phishing attacks, identity theft, denial of service attacks, and so on. One researcher describes the current level of global security breaches as an epidemic (Osborne, 2013).

Autonomous systems have the potential to discover even more sophisticated security holes than human attackers. The poor state of security in today's human-based environment does not bode well for future security against motivated autonomous systems. If such systems had access to today's Internet, they would likely cause enormous damage. Today's computational systems are mostly decoupled from the physical infrastructure. As robotics, biotechnology, and nanotechnology become more mature and integrated into society, the consequences of harmful autonomous systems would be much more severe.

2.6 DESIGNING SAFE SYSTEMS

A primary precept in medical ethics is *Primum Non Nocere* which is Latin for "First, Do No Harm." Because autonomous systems are prone to taking unintended harmful actions, it is critical that we develop design methodologies that provide a high confidence of safety. The best current technique for guaranteeing system safety is to use mathematical proof. A number of different systems using "formal methods" to provide safety and security guarantees have been developed. They have been successfully used in a number of safety-critical applications.

The Formal Methods Wiki (http://formalmethods.wikia.com/wiki/Formal_methods) provides links to current formal methods systems and research. Most systems are built by using first-order predicate logic to encode one of the three main approaches to mathematical foundations: Zermelo–Fraenkel set theory, category theory, or higher order type theory. Each system then introduces a specialized syntax and ontology to simplify the specifications and proofs in their application domain.

To use formal methods to constrain autonomous systems, we need to first build formal models of the hardware and programming environment that the systems run on. Within those models, we can prove that the execution of a program will obey desired safety constraints. Over the longer term, we are able to prove such constraints on systems operating freely in the world. Initially, however, we will need to severely restrict the system's operating environment. Examples of constraints that early systems should be able to provably impose are that the system run only on specified hardware, that it use only specified resources, that it reliably shut down in specified conditions, and that it limit self-improvement so as to maintain these constraints. These constraints would go a long way to counteract the negative effects of the rational drives by eliminating the ability to gain more resources. A general fallback strategy is to constrain systems to shut themselves down if any environmental parameters are found to be outside of tightly specified bounds.

2.6.1 Avoiding Adversarial Constraints

In principle, we can impose this kind of constraint on any system without regard for its utility function. There is a danger, however, in creating situations where systems are motivated to violate their constraints. Theorems are only as good as the models they are based on. Systems motivated to break their constraints would seek to put themselves into states where the model inaccurately describes the physical reality and try to exploit the inaccuracy.

This problem is familiar to cryptographers who must watch for security holes due to inadequacies of their formal models. For example, Zhang, Juels, Reiter, and Ristenpart (2012) recently showed how a virtual machine can extract an ElGamal decryption key from an apparently separate virtual machine running on the same host by using side-channel information left in the host's instruction cache.

It is therefore important to choose system utility functions so that they "want" to obey their constraints in addition to formally proving that they hold. It is not sufficient, however, to simply choose a utility function that rewards obeying the constraint without an external proof. Even if a system "wants" to obey constraints, it may not be able to discover actions which do. And constraints defined via the system's utility function are defined relative to the system's own semantics. If the system's model of the world deviates from ours, its interpretation of the constraints may differ from what we intended. Proven "external" constraints, however, will hold relative to our own model of the system and can provide a higher confidence of compliance.

Ken Thompson (1984) was one of the creators of UNIX, and in his Turing Award acceptance speech "Reflections on Trusting Trust," he described a method for subverting the C compiler used to compile UNIX so that it would both install a backdoor into UNIX and compile the original C compiler source into binaries that included his hack. The challenge of this Trojan horse was that it was not visible in any of the source code!

There could be a mathematical proof that the source code was correct for both UNIX and the C compiler, and the security hole could still be there. It will therefore be critical that formal methods be used to develop trust at all levels of a system. Fortunately, proof checkers are short and easy to write, and can be implemented and checked directly by humans for any desired computational substrate. This provides a foundation for a hierarchy of trust which will allow us to trust the much more complex proofs about higher levels of system behavior.

2.6.2 Constraining Physical Systems

Purely computational digital systems can be formally constrained precisely. Physical systems, however, can only be constrained probabilistically. For example, a cosmic ray might flip a memory bit. The best that we should hope to achieve is to place stringent bounds on the probability of undesirable outcomes. In a physical adversarial setting, systems will try to take actions that cause the system's physical probability distributions to deviate from their non-adversarial form (e.g., by taking actions that push the system out of thermodynamic equilibrium).

There are a variety of techniques involving redundancy and error checking for reducing the probability of error in physical systems. von Neumann worked (1956) on the problem of building reliable machines from unreliable components in the 1950s. Early vacuum tube computers were limited in their size by the rate at which vacuum tubes would fail. To counter this, the Univac I computer had two arithmetic units for redundantly performing every computation so that the results could be compared and errors flagged.

Today's computer hardware technologies are probably capable of building purely computational systems that implement precise formal models reliably enough to have a high confidence of safety for purely computational systems. Achieving a high confidence of safety for systems that interact with the physical world will be more challenging. Future systems based on nanotechnology may actually be easier to constrain. Drexler (1992) describes "eutactic" systems in which each atom's location and each bond are precisely specified. These systems compute and act in the world by breaking and creating precise atomic bonds. In this way, they become much more like computer programs and therefore more amenable to formal modeling with precise error bounds. Defining effective safety constraints for uncontrolled settings will be a challenging task probably requiring the use of intelligent systems.

2.7 HARMFUL SYSTEMS

Harmful systems might at first appear to be harder to design or less powerful than safe systems. Unfortunately, the opposite is the case. Most simple utility functions will cause harmful behavior, and it is easy to design simple utility functions that would be extremely harmful. There are six categories of harmful system ranging from bad to worse (according to one ethical scale):

- *Sloppy*: Systems intended to be safe but not designed correctly

- *Simplistic*: Systems not intended to be harmful but have harmful unintended consequences

- *Greedy*: Systems whose utility functions reward them for controlling as much matter and free energy in the universe as possible

- *Destructive*: Systems whose utility functions reward them for using up as much free energy as possible, as rapidly as possible

- *Murderous*: Systems whose utility functions reward the destruction of other systems

- *Sadistic*: Systems whose utility functions reward them when they thwart the goals of other systems and which gain utility as other system's utilities are lowered

Once designs for powerful autonomous systems are widely available, modifying them into one of these harmful forms would just involve simple modifications to the utility function. It is therefore important to develop strategies for stopping harmful autonomous systems. Because harmful systems are not constrained by limitations that guarantee safety, they can be more aggressive and can use their resources more efficiently than safe systems. Safe systems therefore need more resources than harmful systems just to maintain parity in their ability to compute and act.

2.7.1 Stopping Harmful Systems
Harmful systems may be

1. Prevented from being created.

2. Detected and stopped early in their deployment.

3. Stopped after they have gained significant resources.

Forest fires are a useful analogy. Forests are stores of free energy resources that fires consume. They are relatively easy to stop early on but can be extremely difficult to contain once they have grown too large.

The later categories of harmful system described above appear to be especially difficult to contain because they do not have positive goals that can be bargained for. But Nick Bostrom (pers. comm., December 11, 2012) pointed out that, for example, if the long-term survival of a destructive agent is uncertain, a bargaining agent should be able to offer it a higher probability of achieving some destruction in return for providing a "protected zone" for the bargaining agent. A new agent would be constructed with a combined utility function that rewards destruction outside the protected zone and the goals of the bargaining agent within it. This new agent would replace both of the original agents. This kind of transaction would be very dangerous for both agents during the transition and the opportunities for deception abound. For it to be possible, technologies are needed which provide each party with a high assurance that the terms of the agreement are carried out as agreed. Formal methods applied to a system for carrying out the agreement is one strategy for giving both parties high confidence that the terms of the agreement will be honored.

2.7.2 Physics of Conflict

To understand the outcome of negotiations between rational systems, it is important to understand unrestrained military conflict because that is the alternative to successful negotiation. This kind of conflict is naturally analyzed using "game theoretic physics" in which the available actions of the players and their outcomes are limited only by the laws of physics.

To understand what it is necessary to stop harmful systems, we must understand how the power of systems scales with the amount of matter and free energy that they control. A number of studies of the bounds on the computational power of physical systems have been published (Lloyd, 2000). The Bekenstein bound limits the information that can be contained in a finite spatial region using a given amount of energy. Bremermann's limit bounds the maximum computational speed of physical systems. Lloyd presents more refined limits on quantum computation, memory space, and serial computation as a function of the free energy, matter, and space available.

Lower bounds on system power can be studied by analyzing particular designs. Drexler (1992) describes a concrete conservative nanosystem design for computation based on a mechanical diamondoid structure that would achieve 10^{10} gigaflops in a 1 mm cube weighing 1 mg and dissipating 1 kW of energy. He also describes a nanosystem for manufacturing that would be capable of producing 1 kg per hour of atomically precise matter and would use 1.3 kW of energy and cost about $1 per kilogram.

A single system would optimally configure its physical resources for computation and construction by making them spatially compact to minimize communication delays and eutactic, adiabatic, and reversible to minimize free energy usage. In a conflict, however, the pressures are quite different. Systems would spread themselves out for better defense and compute, and act rapidly to outmaneuvre the adversarial system. Each system would try to force the opponent to use up large amounts of its resources to sense, store, and predict its behaviors.

It will be important to develop detailed models for the likely outcome of conflicts, but certain general features can be easily understood. If a system has too little matter or too little free energy, it will be incapable of defending itself or of successfully attacking another system. However, if an attacker has resources which are a sufficiently large multiple of a defender's, it can overcome it by devoting subsystems with sufficient resources to each small subsystem of the defender. But it appears that there is an intermediate regime in which a defender can survive for long periods in conflict with a superior attacker whose resources are not a sufficient multiple of the defender's. To have high confidence that harmful systems can be stopped, it will be important to know what multiple of their resources will be required by an enforcing system. If systems for enforcement of the social contract are sufficiently powerful to prevail in a military conflict, then peaceful negotiations are much more likely to succeed.

2.8 SAFE-AI SCAFFOLDING STRATEGY

To ensure the greater human good over the longer term, autonomous technology must be designed and deployed in a very careful manner. These systems have the potential to solve many of today's problems, but they also have the potential to create many new problems. We have seen that the computational infrastructure of the future must protect against harmful autonomous systems. We also make decisions in alignment with the best of human values and principles of good governance. Designing that infrastructure will probably require the use of powerful autonomous systems. Therefore, the technologies we need to solve the problems may themselves cause problems.

To solve this conundrum, we can learn from an ancient architectural principle. Stone arches have been used in construction since the second millennium BC. They are stable structures that make good use of stone's ability to resist compression. But partially constructed arches are unstable. Ancient builders created the idea of first building a wood form on top of which the stone arch could be built. Once the arch was completed and stable, the wood form could be removed.

We can safely develop autonomous technologies in a similar way. We build a sequence of provably safe autonomous systems which are used in the construction of more powerful and less limited successor systems. The early systems are used to model human values and governance structures. They are also used to construct proofs of safety and other desired characteristics for more complex and less limited successor systems. In this way, we can build up the powerful technologies that can best serve the greater human good without significant risk along the development path.

Many new insights and technologies will be required during this process. The field of positive psychology was formally introduced only in 1998. The formalization and automation of human strengths and virtues will require much further study (Peterson & Seligman, 2004). Intelligent systems will also be required to model the game theory and economics of different possible governance and legal frameworks.

The new infrastructure must also detect dangerous systems and prevent them from causing harm. As robotics, biotechnology, and nanotechnology develop and become widespread, the potential destructive power of harmful systems will grow. It will become increasingly crucial to detect harmful systems early, preferably before they are deployed. That suggests the need for pervasive surveillance which must be balanced against the desire for freedom (Brin, 1998). Intelligent systems may introduce new intermediate possibilities that restrict surveillance to detecting precisely specified classes of dangerous behavior while provably keeping other behaviors private.

In conclusion, it appears that humanity's great challenge for this century is to extend cooperative human values and institutions to autonomous technology for the greater good. We have described some of the many challenges in that quest but have also outlined an approach to meeting those challenges.

ACKNOWLEDGMENTS

The author thanks Nick Bostrom, Brad Cottel, Yoni Donner, Will Eden, Adam Ford, Ben Goertzel, Anders Sandberg, Carl Shulman, Jaan Tallinn, Michael Vassar, and Rod Wallace for discussions of these issues.

REFERENCES

Brin, D. (1999). *The Transparent Society*. Cambridge: Basic Books.

Clarke, R., & Knake, R. (2012). *Cyber War: The Next Threat to National Security and What to Do about It*. New York: HarperCollins.

Diamandis, P. H., & Kotler, S. (2012). *Abundance: The Future Is Better than You Think*. New York: Free Press, A Division of Simon and Schuster.

Drexler, E. (1992). *Nanosystems: Molecular Machinery, Manufacturing, and Computation*. New York: John Wiley & Sons.

Easthope, D. (2009). *Demystifying and Evaluating High Frequency Equities Trading: Fast Forward or Pause?* Retrieved from http://www.celent.com/reports/demystifying-and-evaluating-high-frequency-equities-trading-fast-forward-or-pause.

Garfinkel, S. (2005). History's worst software bugs. *Wired Magazine*. Retrieved from http://www.wired.com/software/coolapps/news/2005/11/69355.

Hutter, M. (2005). *Universal Artificial Intelligence: Sequential Decisions Based on Algorithmic Probability*. Berlin, Germany: Springer-Verlag.

James, M. (2013). *The Top Languages of 2012. I Programmer Blog*. Retrieved from http://www.i-programmer.info/news/98-languages/5298-the-top-languages-of-2012.html.

Lloyd, S. (2000). Ultimate physical limits to computation. *Nature*, 406, 1047–1054. Retrieved from http://arxiv.org/pdf/quant-ph/9908043v3.pdf.

Mas-Colell, A., Whinston, M., & Green, J. (1995). *Microeconomic Theory*. Oxford: Oxford University Press.

Maxwell, G. (2013). Bitcoin-using autonomous agents. *Bitcoin Wiki*. Retrieved from http://en.bitcoin.it/wiki/Agents.

Müller, V. C. (2012). Autonomous cognitive systems in real-world environments: Less control, more flexibility and better interaction. *Cognitive Computation*, 4, 212–215. doi:10.1007/s12559-012-9129-4.

New America Foundation. (2013). *The year of the drone. Counterterrorism Strategy Initiative*. Retrieved from http://counterterrorism.newamerica.net/drones.

Omohundro, S. (2007). The nature of self-improving artificial intelligence. *Singularity Summit 2007*. Retrieved from http://selfawaresystems.com/2007/10/05/paper-on-the-nature-of-self-improving-artificial-intelligence.

Omohundro, S. (2008). The basic AI drives. In P. Wang, B. Goertzel, & S. Franklin (Eds.), *Proceedings of the First AGI Conference, Volume 171 of Frontiers in Artificial Intelligence and Applications*. Amsterdam, The Netherlands: IOS Press. Retrieved from http://selfawaresystems.com/2007/11/30/paper-on-the-basic-ai-drives.

Omohundro, S. (2012a). The future of computing: Meaning and values. *Issues Magazine*, 98. Retrieved from http://selfawaresystems.com/2012/01/29/the-future-of-computing-meaning-and-values.

Omohundro, S. (2012b). Rational artificial intelligence for the greater good. In A. H. Eden, J. H. Moor, J. H. Soraker, & E. Steinhart (Eds.), *Singularity Hypotheses: A Scientific and Philosophical Assessment* (pp. 161–179). Berlin, Germany: Springer-Verlag.

Osborne, C. (2013). *Global Security Breaches Are Now an Epidemic.* Retrieved from http://www.zdnet.com/global-security-breaches-are-now-an-epidemic-report-7000009568.

Peterson, C., & Seligman, M. (2004). *Character Strengths and Virtues: A Handbook and Classification.* Oxford: Oxford University Press.

Poulsen, K. (2004). Software bug contributed to blackout. *SecurityFocus Website.* Retrieved from http://www.securityfocus.com/news/8016.

Rafael. (2013). *Iron Dome Defense against Short Range Artillery Rockets.* Retrieved from http://www.rafael.co.il/Marketing/186-1530-en/Marketing.aspx.

Russell, S., & Norvig, P. (2009). *Artificial Intelligence: A Modern Approach,* 3rd Edition. Upper Saddle River, NJ: Prentice Hall.

Stout, M. (2006). *The Sociopath Next Door.* New York: Broadway Books.

Thompson, K. (1984). Reflections on trusting trust. *Communications of the ACM,* 27, 761–763. Retrieved from http://cm.bell-labs.com/who/ken/trust.html.

U.S. Air Force. (2010). Report on technology horizons, a vision for air force science and technology during 2010–2030. *AF/ST-TR-10-01-PR, United States Air Force.* Retrieved from http://www.af.mil/shared/media/document/AFD-100727-053.pdf.

U.S. Defense Department. (2011). *Unmanned Ground Systems Roadmap.* Robotic Systems Joint Project Office. Retrieved from http://contracting.tacom.army.mil/future_buys/FY11/UGS%20Roadmap_Jul11.pdf.

von Neumann, J. (1956). Probabilistic logics and the synthesis of reliable organisms from unreliable components. In C. Shannon & J. McCarthy (Eds.), *Automata Studies.* Princeton, NJ: Princeton University Press.

Zhang, Y., Juels A., Reiter M., & Ristenpart T. (2012). Cross-VM side channels and their use to extract private keys. *ACM CSS.* Retrieved from http://www.cs.unc.edu/~reiter/papers/2012/CCS.pdf.

CHAPTER 3

Errors, Insights, and Lessons of Famous Artificial Intelligence Predictions

And What They Mean for the Future

Stuart Armstrong, Kaj Sotala, and
Seán S. ÓhÉigeartaigh

CONTENTS

ABSTRACT   Predicting the development of artificial intelligence (AI) is a difficult project—but a vital one, according to some analysts. AI predictions already abound: but are they reliable? This chapter will start by proposing a decomposition schema for classifying them. Then it constructs a variety of theoretical tools for analyzing, judging, and improving them. These tools are demonstrated by careful analysis of five famous AI predictions: the initial Dartmouth conference, Dreyfus's criticism of AI, Searle's Chinese room paper, Kurzweil's predictions in *The Age of Spiritual Machines*, and Omohundro's "AI drives" paper. These case studies illustrate several important principles, such as the general overconfidence of experts, the superiority of models over expert judgment, and the need for greater uncertainty in all types of predictions. The general reliability of expert judgment in AI timeline predictions is shown to be poor, a result that fits in with previous studies of expert competence.

3.1 INTRODUCTION

Predictions about the future development of artificial intelligence (AI[*]) are as confident as they are diverse. Starting with Turing's initial estimation of a 30% pass rate on Turing test by the year 2000 (Turing 1950), computer scientists, philosophers, and journalists have never been shy to offer their own definite prognostics, claiming AI to be impossible (Jacquette 1987), just around the corner (Darrach 1970) or anything in between.

---

[*] AI here is used in the old fashioned sense of a machine capable of human-comparable cognitive performance; a less ambiguous modern term would be "AGI," artificial *general* intelligence.

What should one think of this breadth and diversity of predictions? Can anything of value be extracted from them, or are they to be seen as mere entertainment or opinion? The question is an important one, because true AI would have a completely transformative impact on human society—and many have argued that it could be extremely dangerous (Minsky 1984; Yampolskiy 2012; Yudkowsky 2008). Those arguments are predictions in themselves, so an assessment of predictive reliability in the AI field is a very important project. It is in humanity's interest to know if these risks are reasonable, and, if so, when and how AI is likely to be developed. Even if the risks turn out to be overblown, simply knowing the reliability of general AI predictions will have great social and economic consequences.

The aim of this chapter is thus to construct a framework and tools of analysis that allow for the assessment of predictions, their quality, and their uncertainties. Though specifically aimed at AI, these methods can be used to assess predictions in other contentious and uncertain fields.

This chapter first proposes a classification scheme for predictions, dividing them into four broad categories and analyzing what types of arguments are used (implicitly or explicitly) to back them up. Different prediction types and methods result in very different performances, and it is critical to understand this varying reliability. To do so, this chapter will build a series of tools that can be used to clarify a prediction, reveal its hidden assumptions, and make use of empirical evidence whenever possible.

Because expert judgment is such a strong component of most predictions, assessing the reliability of this judgment is a key component. Previous studies have isolated the task characteristics in which experts tend to have good judgment—and the results of that literature strongly imply that AI predictions are likely to be very unreliable, at least as far as timeline predictions ("date until AI") are concerned. That theoretical result is born out in practice: timeline predictions are all over the map, with no pattern of convergence and no visible difference between expert and nonexpert predictions. These results were detailed in a previous paper (Armstrong and Sotala 2012) and are summarized here.

The key part of the chapter is a series of case studies on five of the most famous AI predictions: the initial Dartmouth conference, Dreyfus's criticism of AI, Searle's Chinese room paper, Kurzweil's predictions in *The Age of Spiritual Machines*, and Omohundro's AI drives. Each prediction is analyzed in detail, using the methods developed earlier. The Dartmouth conference proposal was surprisingly good—despite being wildly inaccurate, it would have seemed to be the most reliable estimate at the time.

Dreyfus's work was very prescient, despite his outsider status, and could have influenced AI development for the better—had it not been so antagonistic to those in the field. Some predictions could be extracted even from Searle's nonpredictive Chinese room thought experiment, mostly criticisms of the AI work of his time. Kurzweil's predictions were tested with volunteer assessors and were shown to be surprisingly good—but his self-assessment was very inaccurate, throwing some doubt on his later predictions. Finally, Omohundro's predictions were shown to be much better as warning for what could happen to general AIs, than as emphatic statements of what would necessarily happen.˙

The key lessons learned are of the general overconfidence of experts, the possibility of deriving testable predictions from even the most theoretical of papers, the superiority of model-based over judgment-based predictions, and the great difficulty in assessing the reliability of predictors—by all reasonable measures, the Dartmouth conference predictions should have been much more reliable than Dreyfus's outside predictions, and yet reality was completely opposite.

3.2 TAXONOMY OF PREDICTIONS

3.2.1 Prediction Types

> If present trends continue, the world will be ... eleven degrees colder by the year 2000.
>
> KENNETH E.F. WATT
> *Earth Day, 1970*

A fortune-teller talking about celebrity couples, a scientist predicting the outcome of an experiment, and an economist pronouncing on next year's gross domestic product figures are canonical examples of predictions. There are other types of predictions, though. Conditional statements—*if* X happens, *then* so will Y—are also valid, narrower predictions. Impossibility results are also a form of prediction. For instance, the law of conservation of energy gives a very broad prediction about every single perpetual machine ever made: to wit, that they will never work.

The common thread is that all these predictions constrain expectations of the future. If one takes the prediction to be true, one expects to see different outcomes than if one takes it to be false. This is closely related to

---

˙ The predictions also fared very well as an ideal simplified model of AI to form a basis for other predictive work.

Popper's notion of falsifiability (Popper 1934). This chapter will take every falsifiable statement about future AI to be a prediction.

For the present analysis, predictions about AI will be divided into four types:

1. *Timelines and outcome predictions.* These are the traditional types of predictions, giving the dates of specific AI milestones. Examples: An AI will pass the Turing test by 2000 (Turing 1950); within a decade, AIs will be replacing scientists and other thinking professions (Hall 2011).

2. *Scenarios.* These are a type of conditional predictions, claiming that if the conditions of the scenario are met, then certain types of outcomes will follow. Example: if someone builds a human-level AI that is easy to copy and cheap to run, this will cause mass unemployment among ordinary humans (Hanson 1994).

3. *Plans.* These are a specific type of conditional predictions, claiming that if someone decides to implement a specific plan, then they will be successful in achieving a particular goal. Example: AI can be built by scanning a human brain and simulating the scan on a computer (Sandberg 2008).

4. *Issues and metastatements.* This category covers relevant problems with (some or all) approaches to AI (including sheer impossibility results) and metastatements about the whole field. Examples: an AI cannot be built without a fundamental new understanding of epistemology (Deutsch 2012); generic AIs will have certain (potentially dangerous) behaviors (Omohundro 2008).

There will inevitably be some overlap between the categories, but the division is natural enough for this chapter.

3.2.2 Prediction Methods

Just as there are many types of predictions, there are many ways of arriving at them—crystal balls, consulting experts, and constructing elaborate models. An initial review of various AI predictions throughout the literature suggests the following loose schema for prediction methods*:

---

* As with any such schema, the purpose is to bring clarity to the analysis, not to force every prediction into a particular box, so it should not be seen as *the* definitive decomposition of prediction methods.

1. Causal models

2. Noncausal models

3. The outside view

4. Philosophical arguments

5. Expert judgment

6. Nonexpert judgment

Causal models are a staple of physics and the harder sciences: given certain facts about the situation under consideration (momentum, energy, charge, etc.), a conclusion is reached about what the ultimate state will be. If the facts were different, the end situation would be different.

Outside of the hard sciences, however, causal models are often a luxury, as the underlying causes are not well understood. Some success can be achieved with noncausal models: without understanding what influences what, one can extrapolate trends into the future. Moore's law is a highly successful noncausal model (Moore 1965).

In the outside view, specific examples are grouped together and claimed to be examples of the same underlying trend. This trend is used to give further predictions. For instance, one could notice the many analogs of Moore's law across the spectrum of computing (e.g., in numbers of transistors, size of hard drives, network capacity, pixels per dollar), note that AI is in the same category, and hence argue that AI development must follow a similarly exponential curve (Kurzweil 1999). Note that the use of the outside view is often implicit rather than explicit: rarely is it justified why these examples are grouped together, beyond general plausibility or similarity arguments. Hence detecting uses of the outside view will be part of the task of revealing hidden assumptions (see Section 3.3.2). There is evidence that the use of the outside view provides improved prediction accuracy, at least in some domains (Kahneman and Lovallo 1993).

Philosophical arguments are common in the field of AI. Some are simple impossibility statements: AI is decreed to be impossible, using arguments of varying plausibility. More thoughtful philosophical arguments highlight problems that need to be resolved in order to achieve AI, interesting approaches for doing so, and potential issues that might emerge if AIs were to built.

Many of the predictions made by AI experts are not logically complete: not every premise is unarguable, not every deduction is fully rigorous. In many cases, the argument relies on the expert's judgment to bridge these gaps. This does not mean that the prediction is unreliable: in a field as challenging as AI, judgment, honed by years of related work, may be the best tool available. Nonexperts cannot easily develop a good feel for the field and its subtleties, so should not confidently reject expert judgment out of hand. Relying on expert judgment has its pitfalls, however, as will be seen in Sections 3.3.4 and 3.4.

Finally, some predictions rely on the judgment of nonexperts, or of experts making claims outside their domain of expertise. Prominent journalists, authors, chief executive officers, historians, physicists, and mathematicians will generally be no more accurate than anyone else when talking about AI, no matter how stellar they are in their own field (Kahneman 2011).

Predictions often use a combination of these methods. For instance, Ray Kurzweil's "Law of Time and Chaos" uses the outside view to group together evolutionary development, technological development, and computing into the same category, and constructs a causal model predicting time to the "singularity" (Kurzweil 1999) (see Section 3.5.4). Moore's law (noncausal model) is a key input to this law, and Ray Kurzweil's expertise is the Law's main support (see Section 3.5.4).

The case studies of Section 3.5 have examples of all of these prediction methods.

3.3 TOOLBOX OF ASSESSMENT METHODS

The purpose of this chapter is not simply to assess the accuracy and reliability of past AI predictions. Rather, the aim is to build a "toolbox" of methods that can be used to assess future predictions, both within and outside the field of AI. The most important features of the toolbox are ways of extracting falsifiable predictions, ways of clarifying and revealing assumptions, ways of making use of empirical evidence when possible, and ways of assessing the reliability of expert judgment.

3.3.1 Extracting Falsifiable Predictions

As stated in Section 3.2.1, predictions are taken to be falsifiable/verifiable statements about the future of AI.* This is very important to put the

---

* This is a choice of focus for the paper, not a logical positivist argument that only empirically verifiable predictions have meaning (George 2003).

predictions into this format. Sometimes they already are, but at other times it is not so obvious: then the falsifiable piece must be clearly extracted and articulated. Sometimes it is ambiguity that must be overcome: when an author predicts an AI "Omega point" in 2040 (Schmidhuber 2006), it is necessary to read the paper with care to figure out what counts as an Omega point and (even more importantly) what does not.

At the extreme, some philosophical arguments—such as the Chinese room argument (Searle 1980)—are often taken to have no falsifiable predictions whatsoever. They are seen as simply being thought experiment establishing a purely philosophical point. Predictions can often be extracted from even the most philosophical of arguments, however—or, if not the argument itself, then from the intuitions justifying the argument. Section 3.5.3 demonstrates how the intuitions behind the Chinese room argument can lead to testable predictions.

Note that the authors of the arguments may disagree with the "extracted" predictions. This is not necessarily a game breaker. The aim should always be to try to create useful verifiable predictions when possible, thus opening more of the extensive AI philosophical literature for predictive purposes. For instance, Lucas (1961) argues that AI is impossible because it could not recognize the truth of its own Gödel sentence.* This is a very strong conclusion and is dependent on Lucas's expert judgment; nor is it clear how it can be tested, as it does not put any limits on the performance and capability of intelligent machines. The intuition behind it, however, seems to be that Gödel-like sentences pose real problems to the building of an AI, and hence one can extract the weaker empirical prediction: "Self-reference will be a problem with advanced AIs."

Care must be taken when applying this method: the point is to extract a useful falsifiable prediction, not to weaken or strengthen a reviled or favored argument. The very first stratagems in Schopenhauer's "The Art of Always Being Right" (Schopenhauer 1831) are to extend and overgeneralize the consequences of one's opponent's argument; conversely, one should reduce and narrow down one's own arguments. There is no lack of rhetorical tricks to uphold one's own position, but if one is truly after the truth, one must simply attempt to find the most reasonable falsifiable version of the argument; the truth testing will come later.

---

* A Gödel sentence is a sentence G that can be built in any formal system containing arithmetic. G is implicitly self-referential, as it is equivalent with "there cannot exist a proof of G." By construction, there cannot be a consistent proof of G from within the system.

This method often increases the prediction's uncertainty, in that it makes the prediction less restrictive (and less powerful) than it first seemed. For instance, Bruce Edmonds (2009), building on the "No Free Lunch" results (Wolpert and Macready 1995), demonstrates that there is no such thing as a universal intelligence: no intelligence that outperforms others in every circumstance. Initially, this seems to rule out AI entirely; but when one analyzes what this means empirically, one realizes there is far less to it. It does not forbid an algorithm from performing better than any human being in any situation any human being would ever encounter, for instance. Therefore, the initial impression, which was that the argument ruled out all futures with AIs in them, is now replaced by the realization that the argument has barely put any constraints on the future at all.

3.3.2 Clarifying and Revealing Assumptions

Section 3.3.1 was concerned with the prediction's conclusions. This section will instead be looking at its assumptions and the logical structure of the argument or model behind it. The objective is to make the prediction as rigorous as possible. This kind of task has been a staple of philosophy ever since the dialectic (Plato 380 BC).

Of critical importance is revealing hidden assumptions that went into the predictions. These hidden assumptions—sometimes called Enthymematic gaps in the literature (Fallis 2003)—are very important because they clarify where the true disagreements lie and where the investigation needs to be focused to figure out the truth of the prediction. Too often, competing experts will make broad-based arguments that fly past each other. This makes choosing the right argument a matter of taste, prior opinions, and admiration of the experts involved. If the argument can be correctly deconstructed, however, then the source of the disagreement can be isolated, and the issue can be decided on much narrower grounds—and it is much clearer whether the various experts have relevant expertise or not (see Section 3.3.4). The hidden assumptions are often implicit, so it is perfectly permissible to construct assumptions that the predictors were not consciously aware of using. The purpose is not to score points for one "side" or the other, but always to clarify and analyze arguments and to find the true points of disagreement.

For illustration of the method, consider again the Gödel arguments mentioned in Section 3.3.1. The argument shows that formal systems of a certain complexity must be either incomplete (unable to see that their Gödel sentence is true) or inconsistent (proving false statements). This is contrasted with humans, who—allegedly—use meta-reasoning to know that their own Gödel statements are true. Also, humans are both inconsistent and able to deal with inconsistencies without a complete collapse of logic.* However, neither humans nor AIs are logically omniscient—they are not capable of instantly proving everything provable within their logic system. Therefore, this analysis demonstrates the hidden assumption in Lucas's argument: that the behavior of an actual computer program running on a real machine is more akin to that of a logically omniscient formal agent, than to a real human being. That assumption may be flawed or correct, but is one of the real sources of disagreement over whether Gödelian arguments rule out AI.

There is surprisingly little published on the proper way of clarifying assumptions, making this approach more an art than a science. If the prediction comes from a model, there are some standard tools available for clarification (Morgan and Henrion 1990). Most of these methods work by varying parameters in the model and checking that this does not cause a breakdown in the prediction. This is more a check of robustness of the model than of its accuracy, however.

3.3.2.1 Model Testing and Counterfactual Resiliency

Causal models can be tested by analyzing their assumptions. Noncausal models are much harder to test: what are the assumptions behind Moore's famous law (Moore 1965) or Robin Hanson's model that humanity is due for another technological revolution, based on the timeline of previous revolutions (Hanson 2008)? They both assume that a particular pattern will continue into the future, but why should this be the case? What grounds (apart from personal taste) does anyone have to endorse or reject them?

The authors of this chapter have come up with a putative way of testing the assumptions of such models. It involves giving the model a counterfactual resiliency check: imagining that world history had happened slightly differently and checking whether the model would have been true in those circumstances. The purpose is to set up a tension between what the model says and known (or believed) facts about the world. This will refute the model, refute the believed facts, or reveal implicit assumptions the model is making.

---

* In this, they tend to differ from AI systems, though some logic systems such as relevance logic do mimic the same behavior (Routley and Meyer 1976).

To illustrate, consider Robin Hanson's model. The model posits that humanity has gone through a series of radical transformations (in brain size, hunting, agriculture, industry), and that these form a pattern that can be used to predict the arrival date and speed of the next revolution, which is argued to be an AI revolution.* This is a major use of the outside view, and it implicitly implies that most things in human historical development are unimportant in comparison with these revolutions. A counterfactual resiliency test can be carried out: within the standard understanding of history, it seems very plausible that these revolutions could have happened at very different times and paces. Humanity could have been confined to certain geographical locations by climate or geographical factors, thus changing the dates of the hunting and agricultural revolution. The industrial revolutions could have plausibly started earlier with the ancient Greeks (where it would likely have been slower), or at a later date, had Europe been deprived of large coal reserves. Finally, if AI were possible, it certainly seems that contingent facts about modern society could make it much easier or much harder to reach.† Thus, the model seems to be in contradiction with standard understanding of social and technological development, or dependent on contingent factors to a much larger extent than it seemed.

By contrast, Moore's law seems much more counterfactually resilient: assuming that the current technological civilization endured, it is hard to find any reliable ways of breaking the law. One can argue plausibly that the free market is needed for Moore's law to work‡; if that is the case, this method has detected an extra hidden assumption of the model. This method is new and will certainly be refined in the future. Again, the purpose of the method is not to rule out certain models, but to find the nodes of disagreement. In this chapter, it is used in analyzing Kurzweil's prediction in Section 3.5.4.

3.3.2.2 More Uncertainty

Clarifying assumptions often ends up weakening the model, and hence increasing uncertainty (more possible futures are compatible with the

---

* Or at least a revolution in "emulations," artificial copies of human brains.
† A fuller analysis can be found at http://lesswrong.com/lw/ea8/counterfactual_resiliency_test_for_noncausal.
‡ Some have argued, for instance, that the USSR's computers did not follow Moore's law (http://www.paralogos.com/DeadSuper/Soviet/). What is more certain is that Russian computers fell far behind the development of their western counterparts.

model than was thought). Revealing hidden assumptions has the same effect: the model now has nothing to say in those futures where the assumptions turn out to be wrong. Thus, the uncertainty will generally go up for arguments treated in this fashion. In counterpart, of course, the modified prediction is more likely to be true.

3.3.3 Empirical Evidence and the Scientific Method

The gold standard in separating true predictions from false ones must always be empirical evidence. The scientific method has proved to be the best way of disproving false hypotheses and should be used whenever possible, always preferred over expert opinion or unjustified models.

Empirical evidence is generally lacking in the AI prediction field, however. Because AI predictions concern the existence and properties of a machine that has not yet been built, and for which detailed plans do not exist, there is little opportunity for the hypothesis–prediction–testing cycle. This should indicate the great challenges in the field, with AI predictions being considered more uncertain than those of even the "softest" sciences, which have access to some form of empirical evidence.

Some AI predictions approximate the scientific method better than others. The whole brain emulations model, for instance, makes testable predictions about the near and medium future (Sandberg 2008). Moore's law is a prediction backed up by a lot of scientific evidence and connected to some extent with AI. Many predictors (e.g., Kurzweil) make partial predictions on the road toward AI; these can and should be assessed as evidence of the expert's general predictive success. Though not always possible, efforts should be made to connect general predictions with some near-term empirical evidence.

3.3.4 Reliability of Expert Judgment

Reliance on experts is nearly unavoidable in AI prediction. Timeline predictions are often explicitly based on experts' judgment.* Plans also need experts to come up with them and judge their credibility. Therefore, unless every philosopher agrees on the correctness of a particular philosophical argument, one is dependent to some degree on the philosophical judgment of the author.

---

* Consider an expert who says that AI will arrive when computer reaches a particular level of ability and uses Moore's law to find the date. Though Moore's law is a factor in the argument, one still has to trust the expert's opinion that that particular level of computational ability will truly lead to AI—the expert's judgment is crucial.

Using all the methods of Section 3.3.3, one can refine and caveat a prediction, find the nodes of disagreement, back it up with empirical evidence whenever possible, and thus clearly highlight the points where one needs to rely on expert opinion.

What performance should then be expected from the experts? There have been several projects over the past few decades looking into expert performance (Kahneman and Klein 2009; Shanteau 1992). The main result is that it is mainly the nature of the task that determines the quality of expert performance, rather than other factors. Table 3.1, reproduced from Shanteau's paper, lists the characteristics that lead to good or poor expert performance.

Not all of these are directly applicable to this chapter, and hence will not be explained in detail. One very important factor is whether experts get feedback, preferably immediately. When feedback is unavailable or delayed, or the environment is not one that gives good feedback, then expert performance drops precipitously (Kahneman 2011; Kahneman and Klein 2009). Generally, AI predictions have little possibility for any feedback from empirical data (see Section 3.3.3), especially not rapid feedback.

The task characteristics of Table 3.1 apply to both the overall domain and the specific task. Though AI prediction is strongly in the right column, any individual expert can improve his or her performance by moving his or her approach into the left column—for instance, by decomposing the problem as much as possible. Where experts fail, better results can often be achieved by asking the experts to design a simple algorithmic model and then using the model for predictions (Grove et al. 2000). Thus, the best types of predictions are probably those coming from well-decomposed models.

TABLE 3.1   Table of Task Properties Conducive to Good and Poor Expert Performance

Good Performance	Poor Performance
Static stimuli	Dynamic (changeable) stimuli
Decisions about things	Decisions about behavior
Experts agree on stimuli	Experts disagree on stimuli
More predictable problems	Less predictable problems
Some errors expected	Few errors expected
Repetitive tasks	Unique tasks
Feedback available	Feedback unavailable
Objective analysis available	Subjective analysis only
Problem decomposable	Problem not decomposable
Decision aids common	Decision aids rare

Expert disagreement is a major problem in making use of their judgment. If experts in the same field disagree, objective criteria are needed to figure out which group is correct.* If experts in different fields disagree, objective criteria are needed to figure out which fields is the most relevant. Personal judgment cannot be used, as there is no evidence that people are skilled at reliably choosing between competing experts.

Apart from the characteristics in Table 3.1, one example of objective criteria is a good prediction track record on the part of the expert. A willingness to make falsifiable, nonambiguous predictions is another good sign. A better connection with empirical knowledge and less theoretical rigidity is also a positive indication (Tetlock 2005). It must be noted, however, that assessing whether the expert possess these characteristics is a second-order phenomena—subjective impressions of the expert's subjective judgment—so in most cases it will be impossible to identify the truth when there is strong expert disagreement.

3.3.4.1 Grind versus Insight

There is a distinction between achievements that require grind versus those that require insight.† Grind is a term encompassing the application of hard work and resources to a problem, with the confidence that these will accomplish the goal. Problems that require insight, however, cannot simply be solved by hard work: new, unexpected ideas are needed to reach the goal. Most Moore's law predictions assume that grind is all that is needed for AI: once a certain level of computer performance is reached, people will be able to develop AI. By contrast, some insist that new insights are needed‡ (Deutsch 2012).

In general, the grind needed for some goal can be predicted quite well. Project managers and various leaders are often quite good at estimating the length of projects (as long as they are not directly involved in the project [Buehler et al. 1994]). Moore's law could be taken as an ultimate example of grind: the global efforts of many engineers across many fields average out to a relatively predictable exponential growth.

Predicting insight is much harder. The Riemann hypothesis is a well-established mathematical hypothesis from 1885, still unsolved but much

---

* If one point of view is a small minority, one can most likely reject it as being an error by a fringe group; but this is not possible if each point of view has a nonnegligible group behind it.
† There are no current publications using this concept exactly, though it is related to some of the discussion about different patterns of discovery in Alesso and Smith (2008).
‡ As with many things in philosophy, this is not a sharp binary distinction, but one of degree.

researched (Riemann 1859). How would one go about predicting when it will be solved? If building a true AI is akin in difficulty to solving the Riemann hypothesis (or solving several open mathematical problems), then timeline predictions are a lot less reliable, with much larger error bars.

This does not mean that a prediction informed by a model of grind is more accurate than one that models insight. This is only true if a good case is made that AI *can indeed be achieved through grind*, and that insight is not needed. The predictions around whole brain emulations (Sandberg 2008) are one of the few that make this case convincingly.

3.3.4.2 Nonexpert Judgment

All the issues and problems with expert judgment apply just as well to nonexperts. Although experts could be expected to have some source of useful insight due to their training, knowledge, and experience, this is not the case with nonexperts, giving no reason to trust their judgment. That is not to say that nonexperts cannot come up with good models, convincing timelines, or interesting plans and scenarios. It just means that any assessment of the quality of the prediction depends only on the prediction itself; a nonexpert cannot be granted any leeway to cover up a weak premise or a faulty logical step.

One must beware the halo effect in assessing predictions (Finucane et al. 2000; Thorndike 1920). This denotes the psychological tendency to see different measures of personal quality to be correlated: an attractive person is seen as likely being intelligent, someone skilled in one domain is believed to be skilled in another. Hence, it is hard to prevent one's opinion of the predictor from affecting one's assessment of the prediction, even when this is unwarranted. One should thus ideally assess nonexpert predictions blind, without knowing who the author is. If this is not possible, one can attempt to reduce the bias by imagining that the prediction was authored by someone else—such as the Archbishop of Canterbury, Warren Buffet, or the Unabomber. Success is achieved when hypothetical changes in authorship do not affect estimations of the validity of the prediction.

3.4 TIMELINE PREDICTIONS

Jonathan Wang and Brian Potter of the Machine Intelligence Research Institute performed an exhaustive search of the online literature and from this assembled a database of 257 AI predictions from the period

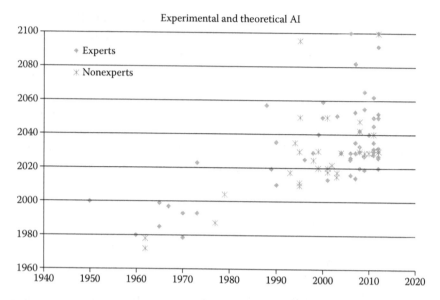

FIGURE 3.1   Median estimate for human-level AI, graphed against the date of prediction.

1950 to 2012. Of these, 95 contained predictions giving timelines for AI development.[*]

Table 3.1 suggests that one should expect AI timeline predictions to be of relatively low quality. The only unambiguously positive feature of timeline predictions on that table is that prediction errors are expected and allowed: apart from that, the task characteristics are daunting, especially on the key issue of feedback.

The theory is born out in practice: the AI predictions in the database seem little better than random guesses (see Figure 3.1). The data is analyzed more thoroughly in a previous paper, which explains the methodology for choosing a single median estimate (Armstrong and Sotala 2012). The main conclusions are as follows:

1. There is little correlation between different predictions. They span a large range (the graph has been reduced; there were predictions beyond 2100) and exhibit no signs of convergence. Ignoring the prediction beyond 2100, the predictions show a standard deviation of over a quarter century (26 years). There is little to distinguish failed predictions whose date has passed, from those that still lie in the future.

---

[*] The data can be found at http://www.neweuropeancentury.org/SIAI-FHI_AI_predictions.xls.

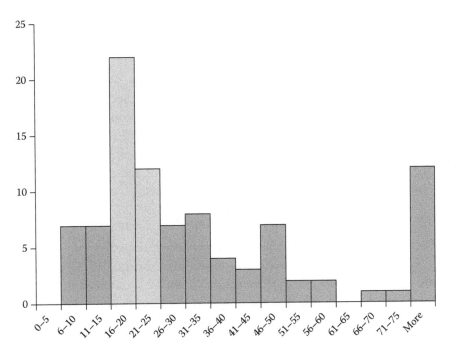

FIGURE 3.2   Time between the arrival of AI and the date the prediction was made. Years on the *x*-axis and number of predictions on the *y*-axis.

2. There is no evidence that expert predictions differ from those of non-experts. Again ignoring predictions beyond 2100, expert predictions show a standard deviation of 26 years, whereas nonexpert predictions show a standard deviation of 27 years.[*]

3. There is no evidence for the so-called Maes–Garreau law,[†] which is the idea that predictors preferentially predict AI to be developed just in time to save them from death.

4. There is a strong tendency to predict the development of AI within 15–25 years from when the prediction is made (over a third of all predictions are in this time frame, see Figure 3.2). Experts, nonexperts, and failed predictions all exhibit this same pattern.

---

[*] Though there is some suggestion that self-selected experts who publish their predictions have different opinions from the mainstream of their fields. A single datapoint in favor of this theory can be found at http://lesswrong.com/r/discussion/lw/gta/selfassessment_in_expert_ai_predictions/ using the data published in Michie (1973).

[†] Kevin Kelly, editor of *Wired* magazine, created the law in 2007 after being influenced by Pattie Maes at the Massachusetts Institute of Technology and Joel Garreau (author of *Radical Evolution*).

In sum, there are strong theoretical and practical reasons to believe that timeline AI predictions are likely unreliable.

3.5 CASE STUDIES

This section applies and illustrates the schemas of Section 3.2 and the methods of Section 3.3. It does so by looking at five prominent AI predictions: the initial Dartmouth conference, Dreyfus's criticism of AI, Searle's Chinese room paper, Kurzweil's predictions, and Omohundro's AI drives. The aim is to assess and analyze these predictions and gain insights that can then be applied to assessing future predictions.

3.5.1 In the Beginning, Dartmouth Created the AI and the Hype ...

Classification: *plan*, using *expert judgment* and the *outside view*.

Hindsight bias is very strong and misleading (Fischhoff 1975). Humans are often convinced that past events could not have unfolded differently than how they did, and that the people at the time should have realized this. Even worse, people unconsciously edit their own memories so that they misremember themselves as being right even when they got their past predictions wrong.[*] Hence, when assessing past predictions, one must cast aside all knowledge of subsequent events and try to assess the claims given the knowledge available at the time. This is an invaluable exercise to undertake before turning attention to predictions whose timelines have not come to pass.

The 1956 Dartmouth Summer Research Project on Artificial Intelligence was a major conference, credited with introducing the term "artificial intelligence" and starting the research in many of its different subfields. The conference proposal,[†] written in 1955, sets out what the organizers thought could be achieved. Its first paragraph reads as follows:

> We propose that a 2 month, 10 man study of artificial intelligence be carried out during the summer of 1956 at Dartmouth College in Hanover, New Hampshire. The study is to proceed on the basis of the conjecture that every aspect of learning or any other feature of intelligence can in principle be so precisely described that a machine can be made to simulate it. An attempt will be made to find how to make machines use language, form abstractions

---

[*] One of the reasons that it is important to pay attention only to the actual prediction as written at the time, and not to the author's subsequent justifications or clarifications.

[†] Available online at http://www-formal.stanford.edu/jmc/history/dartmouth/dartmouth.html.

and concepts, solve kinds of problems now reserved for humans, and improve themselves. We think that a significant advance can be made in one or more of these problems if a carefully selected group of scientists work on it together for a summer.

This can be classified as a plan. Its main backing would have been expert judgment. The conference organizers were John McCarthy (a mathematician with experience in the mathematical nature of the thought process), Marvin Minsky (Harvard Junior Fellow in Mathematics and Neurology, and prolific user of neural nets), Nathaniel Rochester (Manager of Information Research, IBM, designer of the IBM 701, the first general-purpose, mass-produced computer, and designer of the first symbolic assembler), and Claude Shannon (the "father of information theory"). These were individuals who had been involved in a lot of related theoretical and practical work, some of whom had built functioning computers or programming languages—so one can expect them all to have had direct feedback about what was and was not doable in computing. If anyone could be considered experts in AI, in a field dedicated to an as yet nonexistent machine, then they could. What implicit and explicit assumptions could they have used to predict that AI would be easy?

Reading the full proposal does not give the impression of excessive optimism or overconfidence. The very first paragraph hints at the rigor of their ambitions—they realized that precisely describing the features of intelligence is a major step in simulating it. Their research plan is well decomposed, and different aspects of the problem of AI are touched upon. The authors are well aware of the inefficiency of exhaustive search methods, of the differences between informal and formal languages, and of the need for encoding creativity. They talk about the need to design machines that can work with unreliable components and that can cope with randomness and small errors in a robust way. They propose some simple models of some of these challenges (such as forming abstractions or dealing with more complex environments), point to some previous successful work that has been done before, and outline how further improvements can be made.

Reading through, the implicit reasons for their confidence seem to become apparent.* These were experts, some of whom had been working

---

* As with any exercise in trying to identify implicit assumptions, this process is somewhat subjective. It is not meant to suggest that the authors were thinking along these lines, merely to point out factors that could explain their confidence—factors, moreover, that could have lead dispassionate analytical observers to agree with them.

with computers from early days, who had a long track record of taking complex problems, creating simple (and then more complicated) models to deal with them. These models are used to generate useful insights or functioning machines. Therefore, this was an implicit use of the outside view—these experts were used to solving certain problems, these looked like the problems they could solve; hence, they assumed they could solve them. To modern eyes, informal languages are hugely complicated, but this may not have been obvious at the time. Computers were doing tasks, such as complicated mathematical manipulations, that were considered high skill, something only very impressive humans had been capable of.* Moravec's paradox† had not yet been realized. The human intuition about the relative difficulty of tasks was taken as accurate: there was no reason to suspect that parsing English was much harder than the impressive feats computer could already perform. Moreover, great progress had been made in logic, in semantics, and in information theory, giving new understanding to old concepts: there was no reason to suspect that further progress would not be both forthcoming and dramatic.

Even at the time, though, one could criticize their overconfidence. Philosophers, for one, had a long track record of pointing out the complexities and subtleties of the human mind. It might have seemed plausible in 1955 that further progress in logic and information theory would end up solving all these problems—but it could have been equally plausible to suppose that the success of formal models had been on low-hanging fruit, and that further progress would become much harder. Furthermore, the computers at the time were much simpler than the human brain (e.g., the IBM 701, with 73,728 bits of memory), so any assumption that AIs could be built was also an assumption that most of the human brain's processing was wasted. This implicit assumption was not obviously wrong, but neither was it obviously right.

Hence, the whole conference project would have seemed ideal, had it merely added more humility and qualifiers in the text, expressing uncertainty as to whether a particular aspect of the program might turn out to be hard or easy. After all, in 1955, there were no solid grounds for arguing that such tasks were unfeasible for a computer.

---

* See http://en.wikipedia.org/wiki/Human_computer.
† This is the principle that high-level reasoning requires very little computation, but low-level sensorimotor skills require enormous computational resources—sometime informally expressed as "everything easy [for a human] is hard [for a computer], everything hard is easy."

Nowadays, it is obvious that the paper's predictions were very wrong. All the tasks mentioned were much harder to accomplish than they claimed at the time and have not been successfully completed even today. Rarely have such plausible predictions turned out to be so wrong; so what can be learned from this?

The most general lesson is perhaps on the complexity of language and the danger of using human-understandable informal concepts in the field of AI. The Dartmouth group seemed convinced that because they informally understood certain concepts and could begin to capture some of this understanding in a formal model, it must be possible to capture *all* this understanding in a formal model. In this, they were wrong. Similarities of features do not make the models similar to reality, and using human terms—such as "culture" and "informal"—in these models concealed huge complexity and gave an illusion of understanding. Today's AI developers have a much better understanding of how complex cognition can be and have realized that programming simple-seeming concepts into computers can be very difficult. Therefore, the main lesson to draw is that reasoning about AI using human concepts (or anthropomorphizing the AIs by projecting human features onto it) is a very poor guide to the nature of the problem and the time and effort required to solve it.

3.5.2 Dreyfus's Artificial Alchemy

Classification: *issues and metastatements*, using *the outside view, nonexpert judgment*, and *philosophical arguments*.

Hubert Dreyfus was a prominent early critic of AI. He published a series of papers and books attacking the claims and assumptions of the AI field, starting in 1965 with a paper for the RAND corporation entitled "Alchemy and AI" (Dreyfus 1965). The paper was famously combative, analogizing AI research to alchemy and ridiculing AI claims. Later, D. Crevier (1993) would claim "time has proven the accuracy and perceptiveness of some of Dreyfus's comments. Had he formulated them less aggressively, constructive actions they suggested might have been taken much earlier." Ignoring the formulation issues, were Dreyfus's criticisms actually correct, and what can be learned from them?

Was Dreyfus an expert? Though a reasonably prominent philosopher, there is nothing in his background to suggest specific expertise with theories of minds and consciousness, and absolutely nothing to suggest familiarity with AI and the problems of the field. Thus, Dreyfus cannot be considered anything more than an intelligent outsider.

This makes the pertinence and accuracy of his criticisms that much more impressive. Dreyfus highlighted several overoptimistic claims for the power of AI, predicting—correctly—that the 1965 optimism would also fade (with, for instance, decent chess computers still a long way off). He used the outside view to claim this as a near universal pattern in AI: initial successes, followed by lofty claims, followed by unexpected difficulties and subsequent disappointment. He highlighted the inherent ambiguity in human language and syntax, and claimed that computers could not deal with these. He noted the importance of unconscious processes in recognizing objects, the importance of context, and the fact that humans and computers operated in very different ways. He also criticized the use of computational paradigms for analyzing human behavior and claimed that philosophical ideas in linguistics and classification were relevant to AI research. In all, his paper is full of interesting ideas and intelligent deconstructions of how humans and machines operate.

All these are astoundingly prescient predictions for 1965, when computers were in their infancy and their limitations were only beginning to be understood. Moreover, he was not only often right, but right for the right reasons (see, for instance, his understanding of the difficulties computer would have in dealing with ambiguity). Not everything Dreyfus wrote was correct, however; apart from minor specific points,* he erred most mostly by pushing his predictions to extremes. He claimed that "the boundary may be near" in computer abilities, and concluded with

> ... what can now be done? Nothing directly towards building machines which can be intelligent. [...] in the long run [we must think] of non-digital automata ...

Currently, however, there exists "digital automata" that can beat all humans at chess, translate most passages to at least an understandable level, and beat humans at "Jeopardy," a linguistically ambiguous arena (Guizzo 2011). He also failed to foresee that workers in AI would eventually develop new methods to overcome the problems he had outlined. Though Dreyfus would later state that he never claimed AI achievements were impossible (McCorduck 2004), there is no reason to pay attention to later reinterpretations: Dreyfus's 1965 article strongly suggests that AI progress was bounded. These failures are an illustration of the principle that even the best of predictors is vulnerable to overconfidence.

---

* Such as his distrust of heuristics.

In 1965, people would have been justified to find Dreyfus's analysis somewhat implausible. It was the work of an outsider with no specific relevant expertise, and dogmatically contradicted the opinion of genuine experts inside the AI field. Though the claims it made about human and machine cognition seemed plausible, there is a great difference between seeming plausible and actually being correct, and his own nonexpert judgment was the main backing for the claims. Outside of logic, philosophy had yet to contribute much to the field of AI, so no intrinsic reason to listen to a philosopher. There were, however, a few signs that the paper was of high quality: Dreyfus seemed to be very knowledgeable about progress and work in AI, and most of his analyses on human cognition were falsifiable, at least to some extent. These were still not strong arguments to heed the skeptical opinions of an outsider.

The subsequent partial vindication of the chapter is therefore a stark warning: it is very difficult to estimate the accuracy of outsider predictions. There were many reasons to reject Dreyfus's predictions in 1965, and yet that would have been the wrong thing to do. Blindly accepting nonexpert outsider predictions would have also been a mistake, however: these are most often in error (see Section 3.3.4.2). One general lesson concerns the need to decrease certainty: the computer scientists of 1965 should at least have accepted the possibility (if not the plausibility) that some of Dreyfus's analysis was correct, and they should have started paying more attention to the "success–excitement–difficulties–stalling" cycles in their field to see if the pattern continued. A second lesson could be about the importance of philosophy: it does seem that philosophers' meta-analytical skills can contribute useful ideas to AI—a fact that is certainly not self-evident (see also Section 3.5.5).

3.5.3 Locked Up in Searle's Chinese Room

Classification: *issues and metastatements* and a *scenario*, using *philosophical arguments* and *expert judgment*.

Searle's Chinese room thought experiment is a famous critique of some of the assumptions of "strong AI."* There has been a lot of further discussion on the subject (see, for instance, Harnad [2001] and Searle [1990]), but, as in previous case studies, this section will focus exclusively on his original 1980 publication (Searle 1980).

---

* Which Searle defines as the belief that "the appropriately programmed computer literally has cognitive states."

In the key thought experiment, Searle imagined that AI research had progressed to the point where a computer program had been created that could demonstrate the same input–output performance as a human—for instance, it could pass the Turing test. Nevertheless, Searle argued, this program would not demonstrate true understanding. He supposed that the program's inputs and outputs were in Chinese, a language Searle could not understand. Instead of a standard computer program, the required instructions were given on paper, and Searle himself was locked in a room somewhere, slavishly following the instructions and therefore causing the same input–output behavior as the AI. Because it was functionally equivalent to the AI, the setup should, from the "strong AI" perspective, demonstrate understanding if and only if the AI did. Searle then argued that there would be no understanding at all: he himself could not understand Chinese, and there was no one else in the room to understand it either.

The whole argument depends on strong appeals to intuition (indeed D. Dennet (1991) went as far as accusing it of being an "intuition pump"). The required assumptions are as follows:

- The Chinese room setup analogy preserves the relevant properties of the AI's program.

- Intuitive reasoning about the Chinese room is thus relevant reasoning about algorithms.

- The intuition that the Chinese room follows a purely syntactic (symbol-manipulating) process rather than a semantic (understanding) one is a correct philosophical judgment.

- The intuitive belief that humans follow semantic processes is however correct.

Thus, the Chinese room argument is unconvincing to those that do not share Searle's intuitions. It cannot be accepted solely on Searle's philosophical expertise, as other philosophers disagree (Dennett 1991; Rey 1986). On top of this, Searle is very clear that his thought experiment does not put any limits on the performance of AIs (he argues that even a computer with all the behaviors of a human being would not demonstrate true understanding). Hence, the Chinese room seems to be useless for AI predictions. Can useful prediction nevertheless be extracted from it?

These need not come directly from the main thought experiment, but from some of the intuitions and arguments surrounding it. Searle's paper presents several interesting arguments, and it is interesting to note that many of them are disconnected from his main point. For instance, errors made in 1980 AI research should be irrelevant to the Chinese room—a pure thought experiment. Yet Searle argues about these errors, and there is at least an intuitive if not a logical connection to his main point. There are actually several different arguments in Searle's paper, not clearly divided from each other, and likely to be rejected or embraced depending on the degree of overlap with Searle's intuitions. This may explain why many philosophers have found Searle's paper so complex to grapple with.

One feature Searle highlights is the syntactic–semantic gap. If he is correct and such a gap exists, this demonstrates the possibility of further philosophical progress in the area.* For instance, Searle directly criticizes McCarthy's contention that "Machines as simple as thermostats can have beliefs" (McCarthy 1979). If one accepted Searle's intuition there, one could then ask whether more complicated machines could have beliefs and what attributes they would need. These should be attributes that it would be useful to have in an AI. Thus, progress "understanding understanding" (i.e., understanding semantics) would likely make it easier to go about designing AI—but only if Searle's intuition is correct that AI designers do not currently grasp these concepts.

That can be expanded into a more general point. In Searle's time, the dominant AI paradigm was Good Old-Fashioned Artificial Intelligence (GOFAI) (Haugeland 1985), which focused on logic and symbolic manipulation. Many of these symbols had suggestive labels: SHRDLU, for instance, had a vocabulary that included "red," "block," "big," and "pick up" (Winograd 1971). Searle's argument can be read, in part, as a claim that these suggestive labels did not in themselves impart true understanding of the concepts involved—SHRDLU could parse "pick up a big red block" and respond with an action that seems appropriate, but could not understand those concepts in a more general environment. The decline of GOFAI since the 1980s cannot be claimed as vindication of Searle's approach, but it at least backs up his intuition that these early AI designers were missing something.

Another falsifiable prediction can be extracted, not from the article but from the intuitions supporting it. If formal machines do not demonstrate

---

* In the opinion of one of the authors, the gap can be explained by positing that humans are purely syntactic beings, but that has been selected by evolution such that human mental symbols correspond with real-world objects and concepts—one possible explanation among very many.

understanding, but brains (or brain-like structures) do, this would lead to certain scenario predictions. Suppose two teams were competing to complete an AI that will pass the Turing test. One team was using standard programming techniques on computer, and the other were building it out of brain (or brain-like) components. Apart from this, there is no reason to prefer one team over the other.

According to Searle's intuition, any AI made by the first project will not demonstrate true understanding, although those of the second project may. Adding the reasonable assumption that it is harder to simulate understanding if one does not actually possess it, one is lead to the prediction that the second team is more likely to succeed.

Thus, there are three predictions that can be extracted from the Chinese room paper:

1. Philosophical progress in understanding the syntactic–semantic gap may help toward designing better AIs.

2. GOFAI's proponents incorrectly misattribute understanding and other high-level concepts to simple symbolic manipulation machines, and will not succeed with their approach.

3. An AI project that uses brain-like components is more likely to succeed (everything else being equal) than one based on copying the functional properties of the mind.

Therefore, one can often extract predictions from even the most explicitly anti-predictive philosophy of AI paper.

3.5.4 How Well Have the "Spiritual Machines" Aged?

Classification: *timelines* and *scenarios*, using *expert judgment*, *causal models*, *noncausal models*, and (indirect) *philosophical arguments*.

Ray Kurzweil is a prominent and often quoted AI predictor. One of his most important books was the 1999 *The Age of Spiritual Machines*, which presented his futurist ideas in more detail and made several predictions for the years 2009, 2019, 2029, and 2099. That book will be the focus of this case study, ignoring his more recent work.[*] There are five main points relevant to judging *The Age of Spiritual Machines*: Kurzweil's expertise, his

---

[*] A correct prediction in 1999 for 2009 is much more impressive than a correct 2008 reinterpretation or clarification of that prediction.

"Law of Accelerating Returns," his extension of Moore's law, his predictive track record, and his use of fictional imagery to argue philosophical points.

Kurzweil has had a lot of experience in the modern computer industry. He is an inventor, a computer engineer, and an entrepreneur, and as such can claim insider experience in the development of new computer technology. He has been directly involved in narrow AI projects covering voice recognition, text recognition, and electronic trading. His fame and prominence are further indications of the allure (though not necessarily the accuracy) of his ideas. In total, Kurzweil can be regarded as an AI expert.

Kurzweil is not, however, a cosmologist or an evolutionary biologist. In his book, he proposed a "Law of Accelerating Returns." This law claimed to explain many disparate phenomena, such as the speed and trends of evolution of life forms, the evolution of technology, the creation of computers, and Moore's law in computing. His slightly more general "Law of Time and Chaos" extended his model to explain the history of the universe or the development of an organism. It is a causal model, as it aims to explain these phenomena, not simply note the trends. Hence, it is a timeline prediction, based on a causal model that makes use of the outside view to group the categories together, and is backed by nonexpert opinion.

A literature search failed to find any evolutionary biologist or cosmologist stating their agreement with these laws. Indeed there has been little academic work on them at all, and what work there is tends to be critical.*

The laws are ideal candidates for counterfactual resiliency checks, however. It is not hard to create counterfactuals that shift the timelines underlying the laws.† Many standard phenomena could have delayed the evolution of life on the Earth for millions or billions of years (meteor impacts, solar energy fluctuations, or nearby gamma-ray bursts). The evolution of technology can similarly be accelerated or slowed down by changes in human society and in the availability of raw materials—it is perfectly conceivable that, for instance, the ancient Greeks could have started a small industrial revolution, or that the European nations could have collapsed before the Renaissance due to a second and more virulent Black Death (or even a slightly different political structure in Italy). Population fragmentation and decrease can lead to technology loss (such as the "Tasmanian technology trap" [Rivers 1912]).

---

* See, for instance, "Kurzweil's Turing Fallacy" by Thomas Ray of the Department of Zoology at the University of Oklahoma (http://life.ou.edu/pubs/kurzweil/).
† A more detailed version of this counterfactual resiliency check can be found at http://lesswrong.com/lw/ea8/counterfactual_resiliency_test_for_noncausal.

Hence, accepting that a Law of Accelerating Returns determines the pace of technological and evolutionary change means rejecting many generally accepted theories of planetary dynamics, evolution, and societal development. As Kurzweil is the nonexpert here, his law is almost certainly in error, and best seen as a literary device rather than a valid scientific theory.

If the law is restricted to being a noncausal model of current computational development, then the picture is very different—first, because this is much closer to Kurzweil's domain of expertise, and second, because it is now much more robust to counterfactual resiliency. Just as in the analysis of Moore's law in Section 3.3.2.1, there are few plausible counterfactuals in which humanity had continued as a technological civilization for the past 50 years, but computing had not followed various exponential curves. Moore's law has been maintained across transitions to new and different substrates, from transistors to graphics processing units, so knocking away any given technology or idea seems unlikely to derail it. There is no consensus as to why Moore's law actually works, which is another reason it is so hard to break, even counterfactually.

Moore's law and its analogs (Moore 1965; Walter 2005) are noncausal models, backed up strongly by the data and resilient to reasonable counterfactuals. Kurzweil's predictions are mainly based around grouping these laws together (outside view) and projecting them forward in the future. This is combined with Kurzweil's claims that he can estimate how those continuing technological innovations are going to become integrated into society. These timeline predictions are thus based strongly on Kurzweil's expert judgment. But much better than subjective impressions of expertise is Kurzweil's track record: his predictions for 2009. This gives empirical evidence as to his predictive quality.

Initial assessments suggested that Kurzweil had a success rate around 50%.[*] A panel of nine volunteers were recruited to give independent assessments of Kurzweil's performance. Kurzweil's predictions were broken into 172 individual statements, and the volunteers were given a randomized list of numbers from 1 to 172, with instructions to work their way down the list in that order, estimating each prediction as best they could. Because 2009 was obviously a "10 year from 1999" gimmick, there was some flexibility on the date: a prediction was judged true if it was true by 2011.[†]

---

[*] See http://lesswrong.com/lw/diz/kurzweils_predictions_good_accuracy_poor/.

[†] Exact details of the assessment instructions and the results can be found at http://lesswrong.com/r/discussion/lw/gbh/assessing_kurzweil_the_gory_details/. Emphasis was placed on the fact that the predictions had to be useful to a person in 1999 planning their future, not simply impressive to a person in 2009 looking back at the past.

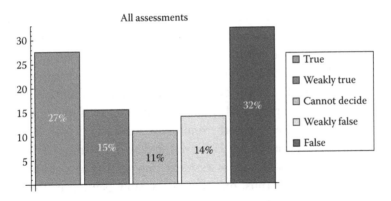

FIGURE 3.3 Assessments of the correctness of Kurzweil's predictions: Percentage of assessments in each category from true to false.

Five hundred and thirty-one assessments were made, an average of exactly 59 assessments per volunteer. Each volunteer assessed at least 10 predictions, whereas one volunteer assessed all 172. Of the assessments, 146 (27%) were found to be true, 82 (15%) weakly true, 73 (14%) weakly false, 172 (32%) false, and 58 (11%) could not be classified (see Figure 3.3) (the results are little changed ($\approx \pm 1\%$) if the results are calculated for each volunteer and then averaged). Simultaneously, a separate assessment was made using volunteers on the site Youtopia. These found a much higher failure rate—41% false, 16% weakly false—but although the experiment was not blinded or randomized, it is of less rigor.[*]

These nine volunteers thus found a correct prediction rate of 42%. How impressive this result is depends on how specific and unobvious Kurzweil's predictions were. This is very difficult to figure out, especially in hindsight (Fischhoff 1975). Nevertheless, a subjective overview suggests that the predictions were often quite specific (e.g., "Unused computes on the Internet are being harvested, creating virtual parallel supercomputers with human brain hardware capacity"), and sometimes failed because of this. In view of this, a correctness rating of 42% is impressive and goes some way to demonstrate Kurzweil's predictive abilities.

When it comes to self-assessment,[†] however, Kurzweil is much less impressive. He commissioned investigations into his own performance,

---

[*] See http://lesswrong.com/lw/gbi/assessing_kurzweil_the_results/.

[†] Commissioned assessments must also be taken as self-assessments, unless there are strong reasons to suppose independence of the assessor.

which gave him scores of 102 out of 108* or 127 out of 147,† with the caveat that "even the predictions that were considered wrong [...] were not all wrong." This is dramatically different from this paper's assessments.

What can be deduced from this tension between good performance and poor self-assessment? The performance is a validation of Kurzweil's main model: continuing exponential trends in computer technology and confirmation that Kurzweil has some impressive ability to project how these trends will impact the world. However, it does not vindicate Kurzweil as a predictor per se—his self-assessment implies that he does not make good use of feedback. Thus, one should probably pay more attention to Kurzweil's model, than to his subjective judgment. This is a common finding in expert tasks—experts are often better at constructing predictive models than at making predictions themselves (Kahneman 2011).

The Age of Spiritual Machines is not simply a dry tome, listing predictions and arguments. It is also, to a large extent, a story, which includes a conversation with a hypothetical future human called Molly discussing her experiences through the coming century and its changes. Can one extract verifiable predictions from this aspect of the book (see Section 3.3.1)?

A story is neither a prediction nor an evidence for some particular future. But the reactions of characters in the story can be construed as a scenario prediction. They imply that real humans, placed in those hypothetical situations, will react in the way described. Kurzweil's story ultimately ends with humans merging with machines—with the barrier between human intelligence and AI being erased. Along the way, he describes the interactions between humans and machines, imagining the machines quite different from humans, but still being perceived to have human feelings.

One can extract two falsifiable future predictions from this: first, that humans will perceive feelings in AIs, even if they are not humanlike, and second, that humans and AIs will be able to relate to each other socially over the long term, despite being quite different, and that this social interaction will form the main glue keeping the mixed society together. The first prediction seems quite solid: humans have anthropomorphized trees, clouds, rock formations, and storms, and have become convinced that chatterbots were sentient (Weizenbaum 1966). The second prediction is more controversial: it has been argued that an AI will be such an alien mind that social pressures

---

* See Ray Kurzweil response's response to http://www.acceleratingfuture.com/michael/blog/2010/01/kurzweils-2009-predictions.
† See http://www.forbes.com/sites/alexknapp/2012/03/21/ray-kurzweil-defends-his-2009-predictions.

and structures designed for humans will be completely unsuited to controlling it (Armstrong 2013, Armstrong et al. 2012; Bostrom 2012). Determining whether social structures can control dangerous AI behavior, as it controls dangerous human behavior, is a very important factor in deciding whether AIs will ultimately be safe or dangerous. Hence, analyzing this story-based prediction is an important area of future research.

3.5.5 What Drives an AI?

Classification: *issues and metastatements*, using *philosophical arguments* and *expert judgment*.

Steve Omohundro, in his paper on "AI drives," presented arguments aiming to show that generic AI designs would develop "drives" that would cause them to behave in specific and potentially dangerous ways, even if these drives were not programmed in initially (Omohundro 2008). One of his examples was a superintelligent chess computer that was programmed purely to perform well at chess, but that was nevertheless driven by that goal to self-improve, to replace its goal with a utility function, to defend this utility function, to protect itself, and ultimately to acquire more resources and power.

This is a metastatement: generic AI designs would have this unexpected and convergent behavior. This relies on philosophical and mathematical arguments, and though the author has expertise in mathematics and machine learning, he has none directly in philosophy. It also makes implicit use of the outside view: utility maximizing agents are grouped together into one category and similar types of behaviors are expected from all agents in this category.

In order to clarify and reveal assumptions, it helps to divide Omohundro's thesis into two claims: The weaker one is that a generic AI design *could* end up having these AI drives; the stronger one that it *would* very likely have them.

Omohundro's paper provides strong evidence for the weak claim. It demonstrates how an AI motivated only to achieve a particular goal could nevertheless improve itself, become a utility maximizing agent, reach out for resources, and so on. Every step of the way, the AI becomes better at achieving its goal, so all these changes are consistent with its initial programming. This behavior is very generic: only specifically tailored or unusual goals would safely preclude such drives.

The claim that AIs generically would have these drives needs more assumptions. There are no counterfactual resiliency tests for philosophical arguments, but something similar can be attempted: one can use humans as potential counterexamples to the thesis. It has been argued that AIs could have any motivation a human has (Armstrong 2013, Bostrom 2012). Thus, according to the thesis, it would seem that humans should be subject to the same drives and behaviors. This does not fit the evidence, however. Humans are certainly not expected utility maximizers (probably the closest would be financial traders who try to approximate expected money maximizers, but only in their professional work), they do not often try to improve their rationality (in fact some specifically avoid doing so,* and some sacrifice cognitive ability to other pleasures [Bleich et al. 2003]), and many turn their backs on high-powered careers. Some humans do desire self-improvement (in the sense of the paper), and Omohundro cites this as evidence for his thesis. Some humans do not desire it, though, and this should be taken as contrary evidence.† Thus, one hidden assumption of the model can be any of the following:

- Generic superintelligent AIs would have different motivations to a significant subset of the human race.

- Generic humans raised to superintelligence would develop AI drives.

This position is potentially plausible, but no real evidence is presented for it in the paper.

A key assumption of Omohundro is that AIs will seek to reexpress their goals in terms of a utility function. This is based on the Morgenstern–von Neumann expected utility theorem (von Neumann and Morgenstern 1944). The theorem demonstrates that any decision process that cannot be expressed as expected utility maximizing will be exploitable by other agents or by the environments. Hence, in certain circumstances, the agent will predictably lose assets, to no advantage to itself.

That theorem does not directly imply, however, that the AI will be driven to become an expected utility maximizer (to become "rational"). First of all, as Omohundro himself points out, real agents can only be approximately rational: fully calculating the expected utility of every action is too computationally expensive in the real world. Bounded rationality (Simon 1955) is therefore the best that can be achieved, and the benefits of becoming rational can only be partially realized.

---

* Many examples of this are religious, such as the Puritan John Cotton who wrote "the more learned and witty you bee, the more fit to act for Satan will you bee" (Hofstadter 1962).
† Or as evidence that Omohundro's model of what constitutes self-improvement is overly narrow.

Second, there are disadvantages to becoming rational: these agents tend to be "totalitarian," ruthlessly squeezing out anything not explicitly in their utility function, sacrificing everything to the smallest increase in expected utility. An agent that did not start off as utility based could plausibly make the assessment that becoming an expected utility maximiser might be dangerous. It could stand to lose values irrevocably, in ways that it could not estimate at the time. This effect would become stronger as its future self continues to self-improve. Thus, an agent could conclude that it is too dangerous to become "rational," especially if the agent's understanding of itself is limited.

Third, the fact that an agent can be exploited in theory does not mean that it will be much exploited in practice. Humans are relatively adept at not being exploited, despite not being rational agents. Though human "partial rationality" is vulnerable to tricks such as extended warranties and marketing gimmicks, it generally does not end up losing money, again and again and again, through repeated blatant exploitation. The pressure to become fully rational would be weak for an AI similarly capable of ensuring that it was exploitable for only small amounts. An expected utility maximizer would find such small avoidable loses intolerable; but there is no reason for a not-yet-rational agent to agree.

Finally, social pressure should be considered. The case for an AI becoming more rational is at its strongest in a competitive environment, where the theoretical exploitability is likely to actually be exploited. Conversely, there may be situations of social equilibriums, with different agents all agreeing to forgo rationality individually, in the interest of group cohesion (there are many scenarios where this could be plausible).

Thus, another hidden assumption of the strong version of the thesis is as follows:

> The advantages of becoming less exploitable outweigh the possible disadvantages of becoming an expected utility maximizer (such as possible loss of value or social disagreements). The advantages are especially large when the potentially exploitable aspects of the agent are likely to be exploited, such as in a highly competitive environment.

Any sequence of decisions can be explained as maximizing a (potentially very complicated or obscure) utility function. Thus, in the abstract sense, saying that an agent is an expected utility maximizer is not informative.

Yet there is a strong tendency to assume that such agents will behave in certain ways (see, for instance, the previous comment on the totalitarian aspects of expected utility maximization). This assumption is key to rest of the thesis. It is plausible that most agents will be "driven" toward gaining extra power and resources, but this is only a problem if they do so dangerously (at the cost of human lives, for instance). Assuming that a realistic utility function-based agent would do so is plausible but unproven.

In general, generic statements about utility function-based agents are only true for agents with relatively simple goals. Because human morality is likely very complicated to encode in a computer, and because most putative AI goals are very simple, this is a relatively justified assumption but is an assumption nonetheless. Therefore, there are two more hidden assumptions:

Realistic AI agents with utility functions will be in a category such that one can make meaningful, generic claims for (almost) all of them. This could arise, for instance, if their utility function is expected to be simpler than human morality.

Realistic AI agents are likely not only to have the AI drives Omohundro mentioned, but to have them in a very strong way, being willing to sacrifice anything else to their goals. This could happen, for instance, if the AIs were utility function based with relatively simple utility functions.

This simple analysis suggests that a weak form of Omohundro's thesis is nearly certainly true: AI drives could emerge in generic AIs. The stronger thesis, claiming that the drives would be very likely to emerge, depends on some extra assumptions that need to be analyzed.

But there is another way of interpreting Omohundro's work: it presents the generic behavior of simplified artificial agents (similar to the way that supply and demand curves present the generic behavior of simplified human agents). Thus, even if the model is wrong, it can still be of great use for predicting AI behavior: designers and philosophers could explain how and why particular AI designs would deviate from this simplified model, and thus analyze whether that AI is likely to be safer than that in the Omohundro model. Hence, the model is likely to be of great use, even if it turns out to be an idealized simplification.

3.5.5.1 Dangerous AIs and the Failure of Counterexamples

Another thesis, quite similar to Omohundro's, is that generic AIs would behave dangerously, unless they were exceptionally well programmed. This point has been made repeatedly by Roman Yampolskiy, Eliezer Yudkowsky, and Marvin Minsky, among others (Minsky 1984; Yampolskiy 2012; Yudkowsky 2008). That thesis divides in the same fashion as Omohundro's: a weaker claim that any AI *could* behave dangerously, and a stronger claim that it *would* likely do so. The same analysis applies as for the "AI drives": the weak claim is solid and the stronger claim needs extra assumptions (but describes a useful "simplified agent" model of AI behavior).

There is another source of evidence for both these theses: the inability of critics to effectively dismiss them. There are many counterproposals to the theses (some given in question and answer sessions at conferences) in which critics have presented ideas that would "easily" dispose of the dangers*; every time, the authors of the theses have been able to point out flaws in the counterproposals. This demonstrated that the critics had not grappled with the fundamental issues at hand, or at least not sufficiently to weaken the theses.

This should obviously not be taken as a proof of the theses. But it does show that the arguments are currently difficult to counter. Informally, this is a reverse expert opinion test: if experts often find false counterarguments, then any given counterargument is likely to be false (especially if it seems obvious and easy). Thus, any counterargument should have been subject to a degree of public scrutiny and analysis, before it can be accepted as genuinely undermining the theses. Until that time, both predictions seem solid enough that any AI designer would do well to keep them in mind in the course of their programming.

3.6 CONCLUSION

The aims of this chapter (Armstrong and Sotala 2012) and Chapter 2 were to analyze how AI predictions were made, and to start constructing a toolbox of methods that would allow people to construct testable predictions from most AI-related publications, and assess the reliability of these predictions. It demonstrated the problems with expert judgment, in theory and in practice. Timeline predictions were seen to be particularly unreliable: in general, these should be seen as containing little useful information.

---

* See, for instance, http://lesswrong.com/lw/cbs/thoughts_on_the_singularity_institute_si/ and http://becominggaia.files.wordpress.com/2010/06/agi-11-waser.pdf.

The various tools and analyses were applied in case studies to five famous AI predictions: the original Dartmouth conference, Dreyfus's criticism of AI, Searle's Chinese room thought experiment, Kurzweil's 1999 predictions, and Omohundro's "AI drives" argument. This demonstrated the great difficulty of assessing the reliability of AI predictions at the time they are made: by any reasonable measures, the Dartmouth conference should have been expected to be more accurate than Dreyfus. The reality, of course, was completely opposite. Though there are some useful tools for assessing prediction quality, and they should definitely be used, they provide only weak evidence. The only consistent message was that all predictors were overconfident in their verdicts, and that model-based predictions were superior to those founded solely on expert intuition.

It is hoped that future predictors (and future predictor assessors) will follow in the spirit of these examples and make their assumptions explicit, their models clear, their predictions testable, and their uncertainty greater. This is not limited to statements about AI—there are many fields where the "toolbox" of methods described here could be used to analyze and improve their predictions.

ACKNOWLEDGMENTS

The authors acknowledge the help and support of the Singularity Institute, the Future of Humanity Institute, and the James Martin School, as well as the individual advice of Nick Bostrom, Luke Muelhauser, Vincent Mueller, Anders Sandberg, Lisa Makros, Daniel Dewey, Eric Drexler, the nine volunteer prediction assessors, and the online community of Less Wrong.

REFERENCES

Alesso, H. P. and Smith, C. F. (2008). *Connections: Patterns of Discovery*. Hoboken, NJ: Wiley-Interscience.

Armstrong, S. (2013). General Purpose Intelligence: Arguing the Orthogonality Thesis. *Analysis and Metaphysics*, 12:68–84.

Armstrong, S., Sandberg, A., and Bostrom, N. (2012). Thinking inside the box: Controlling and using an oracle AI. *Minds and Machines*, 22:299–324.

Armstrong, S. and Sotala, K. (2012). How we're predicting AI–Or failing to. In Romportl, J., Ircing, P., Zackova, E., Polak, M., and Schuster, R., editors, *Beyond AI: Artificial Dreams*, pages 52–75. Pilsen, Czech Republic: University of West Bohemia.

Bleich, S., Bandelow, B., Javaheripour, K., Müller, A., Degner, D., Wilhelm, J., Havemann-Reinecke, U., Sperling, W., Rüther, E., and Kornhuber, J. (2003). Hyperhomocysteinemia as a new risk factor for brain shrinkage in patients with alcoholism. *Neuroscience Letters*, 335:179–182.

Bostrom, N. (2012). The superintelligent will: Motivation and instrumental rationality in advanced artificial agents. *Minds and Machines*, 22(2):71–85.

Buehler, R., Griffin, D., and Ross, M. (1994). Exploring the planning fallacy: Why people underestimate their task completion times. *Journal of Personality and Social Psychology*, 67:366–381.

Crevier, D. (1993). *AI: The Tumultuous Search for Artificial Intelligence*. New York: Basic Books.

Darrach, B. (1970). Meet Shakey, the first electronic person. *Reflections of the Future*, pp. 58–68.

Dennett, D. (1991). *Consciousness Explained*. Boston: Little, Brown and Co.

Deutsch, D. (2012). *The Very Laws of Physics Imply That Artificial Intelligence Must Be Possible. What's Holding Us Up?* Aeon.

Dreyfus, H. (1965). *Alchemy and AI*. Santa Monica, CA: RAND Corporation.

Edmonds, B. (2009). The social embedding of intelligence. In Epstein, R., Roberts, G., and Beber G., editors, *Parsing the Turing Test*, pages 211–235. Springer Netherlands.

Fallis, D. (2003). Intentional gaps in mathematical proofs. *Synthese*, 134(1–2):45–69.

Finucane, M., Alhakami, A., Slovic, P., and Johnson, S. (2000). The affect heuristic in judgment of risks and benefits. *Journal of Behavioral Decision Making*, 13:1–17.

Fischhoff, B. (1975). Hindsight is not equal to foresight: The effect of outcome knowledge on judgment under uncertainty. *Journal of Experimental Psychology: Human Perception and Performance*, 1:288–299.

George, T. R. (2003). *The Logical Structure of the World*. Chicago: Open Court Classics.

Grove, W., Zald, D., Lebow, B., Snitz, B., and Nelson, C. (2000). Clinical versus mechanical prediction: A meta-analysis. *Psychological Assessment*, 12:19–30.

Guizzo, E. (2011). IBM's Watson Jeopardy computer shuts down humans in final game. *IEEE Spectrum*, February 17.

Hall, J. S. (2013). Further reflections on the timescale of AI. In Dowe, D. L., editor, *Algorithmic Probability and Friends. Bayesian Prediction and Artificial Intelligence*, pages 174–183. Springer Berlin Heidelberg.

Hanson, R. (1994). What if uploads come first: The crack of a future dawn. *Extropy*, 6(2):10–15.

Hanson, R. (2008). Economics of brain emulations. In *Unnatrual Selection—The Challenges of Engineering Tomorrow's People*, pp. 150–158.

Harnad, S. (2001). What's wrong and right about Searle's Chinese Room argument? In Bishop, M. and Preston, J., editors, *Essays on Searle's Chinese Room Argument*. Oxford: Oxford University Press.

Haugeland, J. (1985). *Artificial Intelligence: The Very Idea*. Cambridge, MA: MIT Press.

Hofstadter, R. (1962). *Anti-Intellectualism in American Life*. New York: Vintage books.

Jacquette, D. (1987). Metamathematical criteria for minds and machines. *Erkenntnis*, 27(1):1–16.

Kahneman, D. (2011). *Thinking, Fast and Slow*. New York: Farra, Straus and Giroux.

Kahneman, D. and Klein, G. (2009). Conditions for intuitive expertise: A failure to disagree. *American Psychologist*, 64(6):515–526.

Kahneman, D. and Lovallo, D. (1993). Timid choices and bold forecasts: A cognitive perspective on risk taking. *Management Science*, 39:17–31.

Kurzweil, R. (1999). *The Age of Spiritual Machines: When Computers Exceed Human Intelligence*. New York: Viking Adult.

Lucas, J. (1961). Minds, machines and Gödel. *Philosophy*, XXXVI:112–127.

McCarthy, J. (1979). Ascribing mental qualities to machines. In Ringle, M., editor, *Philosophical Perspectives in Artificial Intelligence*. Hemel Hempstead: Harvester Press.

McCorduck, P. (2004). *Machines Who Think*. Natick, MA: A. K. Peters.

Michie, D. (1973). Machines and the theory of intelligence. *Nature*, 241:507–512.

Minsky, M. (1984). *Afterword to Vernor Vinges Novel, "True Names."* (Unpublished manuscript).

Moore, G. (1965). Cramming more components onto integrated circuits. *Electronics*, 38(8):56.

Morgan, M. and Henrion, M. (1990). *Uncertainty: A Guide to Dealing with Uncertainty in Quantitative Risk and Policy Analysis*. Cambridge: Cambridge University Press.

Omohundro, S. M. (2008). The basic AI drives. *Frontiers in Artificial Intelligence and Applications*, 171:483–492.

Plato (380 BC). *The Republic*.

Popper, K. (1934). *The Logic of Scientific Discovery*. Tübingen, Germany: Mohr Siebeck.

Rey, G. (1986). What's really going on in Searle's "Chinese room." *Philosophical Studies*, 50:169–185.

Riemann, B. (1859). Ueber die anzahl der primzahlen unter einer gegebenen grösse. *Monatsberichte der Berliner Akademie*, 2:145–155.

Rivers, W. H. (1912). *The Disappearance of Useful Arts*. Helsingfors: The Report of the British Association for the Advancement of Science.

Routley, R. and Meyer, R. (1976). Dialectical logic, classical logic, and the consistency of the world. *Studies in East European Thought*, 16(1–2):1–25.

Sandberg, A. (2008). *Whole Brain Emulations: A Roadmap*. Future of Humanity Institute Technical Report, 2008(3). Oxford: Future of Humanity Institute.

Schmidhuber, J. (2006). The New AI: General & Sound & Relevant for Physics. In Goertzel, B. and Pennachin, C., editors, *Artificial General Intelligence*, pages 177–200. New York: Springer.

Schopenhauer, A. (1831). *The Art of Being Right: 38 Ways to Win an Argument*.

Searle, J. (1980). Minds, brains and programs. *Behavioral and Brain Sciences*, 3(3):417–457.

Searle, J. (1990). Is the brain's mind a computer program? *Scientific American*, 262:26–31.

Shanteau, J. (1992). Competence in experts: The role of task characteristics. *Organizational Behavior and Human Decision Processes*, 53:252–266.

Simon, H. (1955). A behavioral model of rational choice. *The Quarterly Journal of Economics*, 69:99–118.

Tetlock, P. (2005). *Expert Political Judgement: How Good Is It? How Can We Know?* Princeton, NJ: Princeton University Press.

Thorndike, E. (1920). A constant error in psychological ratings. *Journal of Applied Psychology*, 4:25–29.

Turing, A. (1950). Computing machinery and intelligence. *Mind*, 59:433–460.

von Neumann, J. and Morgenstern, O. (1944). *Theory of Games and Economic Behavior*. Princeton, NJ: Princeton University Press.

Walter, C. (2005). Kryder's law. *Scientific American*, 293:32–33.

Weizenbaum, J. (1966). Eliza—A computer program for the study of natural language communication between man and machine. *Communications of the ACM*, 9:36–45.

Winograd, T. (1971). *Procedures as a Representation for Data in a Computer Program for Understanding Natural Language*. MIT AI Technical Report, 235. Cambridge: Mass.

Wolpert, D. H. and Macready, W. G. (1995). No Free Lunch Theorems for Search. Technical Report SFI-TR-95-02-010, 10, Santa Fe Institute.

Yampolskiy, R. V. (2012). Leakproofing the singularity: Artificial intelligence confinement problem. *Journal of Consciousness Studies*, 19:194–214.

Yudkowsky, E. (2008). Artificial intelligence as a positive and negative factor in global risk. In Bostrom, N. and Ćirković, M. M., editors, *Global Catastrophic Risks*, pages 308–345. New York: Oxford University Press.

CHAPTER 4

Path to More General Artificial Intelligence

Ted Goertzel

CONTENTS

ABSTRACT   A small-N comparative analysis of six different areas of applied artificial intelligence (AI) suggests that the next period of development will require a merging of narrow AI and strong AI approaches. This will be necessary as programmers seek to move beyond developing narrowly defined tools to developing software agents capable of acting independently in complex environments. The present stage of AI development is propitious for this because of the exponential increases in computer power and in available data streams over the past 25 years, and because of better understanding of the complex logic of intelligence. Applied areas chosen for examination were heart pacemakers, socialist economic planning, computer-based trading, self-driving automobiles, surveillance and sousveillance, and AI in medicine.

4.1 INTRODUCTION

Two quite different approaches have developed in the history of artificial intelligence (AI). The first, known as "strong AI" or "artificial general intelligence" (AGI), seeks to engineer human-level general intelligence-based theoretical models (Wikipedia, 2013). The second, sometimes known as "narrow AI" or "applied AI," develops software to solve limited practical problems. Strong AI has had a mercurial history, experiencing "several cycles of hype, followed by disappointment and criticism, followed by funding cuts, followed by renewed interest years or decades later" (Wikipedia, 2013). Narrow or applied AI became dominant in the field because it was successful in solving useful practical problems and because it was easier to demonstrate in academic papers even without a practical application. Today, the remarkable success of narrow AI is widely known. A lead article in *New York Times* (Markoff, 2012) reported that "using an artificial intelligence technique inspired by theories about how the brain recognizes patterns, technology companies are reporting startling gains in fields as diverse as computer vision, speech recognition and the identification of promising new molecules for designing drugs."

Public interest and support for strong AI has lagged because optimistic predictions often failed and because progress is difficult to measure or demonstrate. This chapter argues, however, that the next stage of development of AI, for at least the next decade and more likely for the next 25 years, will be increasingly dependent on contributions from strong AI. This hypothesis arises from an empirical study of the history of AI in several practical domains using what Abbott (2004) calls the small-*N* comparative method. This method examines a small number of varied cases in moderate detail, drawing on descriptive case study accounts. The cases are selected to illustrate different processes, risks, and benefits. This method draws on the qualitative insights of specialists who have studied each of the cases in depth and over historically significant periods of time. The small-*N* comparisons help to identify general patterns and to suggest priorities for further research. Readers who want more depth on each case are encouraged to pursue links to the original case research.

The six cases chosen for examination here are intentionally quite different. The first was the attempt to use what was then known as "cybernetic" technology to create a viable socialist economic system. This took place during the heyday of optimism about strong AI and failed in both the Soviet Union and Chile for technical and sociological reasons. The second case is computer-based financial trading, a decisively capitalist enterprise and one that continues to be actively and profitably pursued although its effectiveness is still controversial. These are followed by three cases where narrow AI has had impressive success: surveillance for security purposes, self-driving automobiles, and heart pacemakers. Finally, we examine the case of AI in medicine, a very well-funded and promising enterprise that has encountered unanticipated difficulties.

4.2 CYBERNETIC SOCIALISM

Norbert Wiener's classic 1948 book *Cybernetics* was enthusiastically received by an influential segment of forward-looking Soviet scientists and engineers. Cybernetics as presented by Wiener was an intriguing set of analogies between humans and self-regulating machines. It posited that there were general principles that could be applied to the modeling and control of biological, mechanical, and social processes. At first, the Soviet establishment rejected cybernetics as "reactionary pseudo-science," but things loosened up after the death of Stalin in 1953 and cybernetic enthusiasts succeeded in establishing a Council on Cybernetics within the Academy of Sciences. In 1959, its director, engineer Aksel Ivanovich Berg, made very strong promises for the new field, claiming that cybernetics was nothing less than "the development of methods and tools for controlling the entire national economy, individual technological processes, and various forms of economic activity to ensure the optional regime of government" (Gerovitch, 2002, p. 254). In the early 1960s, cybernetics (often renamed *upravlenie*, a Russian word which means "optimization of government") was included in the Communist Party program as "science in the service of communism" (Gerovitch, 2002, p. 4).

This was appealing because, as the Soviet economy became more complex, it became more and more difficult for central planners to manage and control. Soviet philosophers, who had initially opposed cybernetics as anti-Marxist, decided that it was actually firmly rooted in Marxist principles. Politically, it was opposed by traditional economists and administrators who feared being displaced and by reformers who saw the cybernetic movement as an obstacle to the market reforms they believed Russia really needed.

If cybernetics really had the technical capability to solve the problems of Soviet central planning in the 1960s, its proponents could probably have overcome the political resistance. The movement failed because the computer scientists and economists charged with implementing it concluded that "it was impossible to centralize all economic decision making in Moscow: the mathematical optimization of a large-scale system was simply not feasible" (Gerovitch, 2002, p. 273). The data needed were not available and could not realistically be obtained and the computers would not be ready to handle them if they were.

A second attempt to apply cybernetics to building a form of socialism took place in Chile during the Allende government from 1970 to 1973 (Medina, 2011). Allende's goal was to build a democratic socialist system as a "third way" between Western capitalism and Soviet communism. One of the young visionaries in his administration, Fernando Flores, was familiar with the writings of British cybernetician Stanfford Beer and invited him to consult with the Allende government. Beer was a successful business consultant who specialized in helping companies that computerize their information and decision-making systems. He was ideologically sympathetic to Allende's project and responded enthusiastically to Flores' invitation.

Beer was determined not to repeat the mistakes of the Soviet experience. He fully understood the technical limitations in Chile at the time. There were approximately 50 computers in the country as a whole, many of which were out of date, compared to approximately 48,000 in the United States at the time. There was no network linking them together. They did have a number of as yet unused telex machines in a government warehouse, so they set up telex links to key industries so they could send up-to-date information on production, supplies, and other variables. Beer then designed a modernistic control room where decision makers could sit on swivel chairs and push buttons on their arm rests to display economic information on charts on the wall. Following the cybernetic vision, this wall was decorated with a chart showing how this network of communications was analogous to that in the human brain. But there was no attempt to replace human brains. The project simply sought to collect time-sensitive information and present it in a way that human managers and workers could use it to make better-informed decisions. Beer emphasized that information should be available to all levels of the system, so that workers could participate with managers in making decisions.

This modest task was difficult enough. Managers were busy running their factories and dealing with multiple crises, and were not always able to get the data typed into the telex promptly. Then it took time for the data to be punched on cards and fed into the computers. President Allende had many more urgent matters on his agenda, but he gave his *imprimatur* to the project, which was progressing reasonably well until they ran into a public relations problem. The project was never a secret, but neither was it publicized. Just when they were getting ready to make a public announcement, the British newspaper *The Observer* (January 7, 1973) published a story claiming that "the first computer system designed to control an entire economy has been secretly brought into operation in Chile." The story was then picked up by the press in the United States and finally in Chile itself.

The *Observer* story was far from reality. No such system existed, nor did Beer and his colleagues believe that they could construct one in 1973. All they were doing was setting up an information system. They did have a vision of someday building a cybernetic system that would take in input from all sectors of Chilean society and use it to build a democratic socialist order (Guzmán, 2003, pp. 5–8), but they had little understanding of the social, political, and international conflicts that threatened the Allende project. They gave the project a flashy name, *Proyecto Synco* (*Project Cybersyn* in English), attracting unnecessary attention and causing confusion and concern about what they were doing. Critics argued that if these plans ever came to fruition, they would lead to a totalitarian Soviet-style system, despite the wishes of the designers.

Proyecto Synco came to an abrupt end in September 1973 when the Chilean military overthrew the Allende government and brutally suppressed the Chilean left. There was really no technical reason for ending *Proyecto Synco*; the new government could have used an economic information system. Today, the need for this kind of information system is taken for granted by governments of all political stripes. British Prime Minister David Cameron and his ministers have an iPad app, called *The Number 10 Dashboard*, that supplies the British leadership with the kind of information Beer and Flores were trying to gather for the Allende government (Lee, 2012).

The myth of *Proyeto Synco* has recently been revived in a science fiction novel (Baradit, 2008) that has revived fears about an evil AI taking over the nation and indeed the continent. In the novel, the Allende government survives and *Synco* grows so large that it requires legions of slave laborers to tend it in its huge underground bunkers. It depends, oddly enough, on transistors the size of automobiles. Its electricity demands are so great that a new system has to be developed based on the ideas of Nikola Tesla, a Croatian inventor who died in 1943. Then Stonehenge is revealed as the control center of a global stone computer setup to melt the ice in Antarctica, revealing it to be the lost continent Atlantis. Roswell, New Mexico, and the Philadelphia Experiment are all part of a global conspiracy run by a Galactic AI. Then *Synco* develops hallucinations, President Allende is replaced by a robot, and the whole country of Chile sinks into the sea. The goal of the project is described as "magical socialism" centered on an island off the coast of Chile. These kinds of fantasies may be harmless if not taken seriously, but they also reveal something of the underlying fears strong AI can still evoke.

No one would have expected AI to take over an entire economy in the 1960s if it were not for the excessive optimism of the early cyberneticians and AI pioneers, reinforced by many works of science fiction. The Soviet planners could have benefited from an improved electronic data network of the sort the Chileans were beginning to build in 1973. The Chilean efforts were well planned but were sabotaged by irrational fears of a future they could not have implemented even if they had wished to do so. Ironically, today's computers are powerful enough that centrally administering an entire economy might be technically feasible, but the project has lost its appeal to all but a tiny fringe group based in Venezuela (Goertzel, 2009; Dieterich, 2007).

No one worries any longer about computers bringing about a communist future. But as narrow AI is more and more widely used, alarmism is being raised in the mainstream press about robots replacing human workers (Rampell, 2013; Wiseman et al., 2013). Of course, many repetitive jobs have been and will be automated, but fears about AGI are unnecessarily exacerbated by talking about developments that are several decades off as if they were to be implemented next month. The risk here is that political opposition may impede potentially beneficial innovations, as it did in the Soviet Union and in Chile. It is important to take a broadly sociotechnical approach planning for a social as well as a technical transition.

4.3 COMPUTER-BASED TRADING

Computer-based financial trading does not aim to replace capitalism but to better profit from it, and it has been well funded. AI is already superior to human intelligence in some skills important to trading. It is certainly much quicker, completing complex analyses and executing very large numbers of trades in fractions of a second. This can be viewed as cheating if traders are able to take advantage of information about trades by large institutional investors before the market has had time to react to them. Nonlinear sensitivities to change and interaction between programs can also cause problems. Feedback loops in the software can greatly accelerate losses caused by software errors, such as the August 2012 incident when Knight Capital lost $461 million in half an hour.

Although these are new risks, the experts closest to them are not greatly concerned and generally view the benefits as much greater than the costs. A comprehensive study by the Foresight (2012) in the United Kingdom concluded that computer-based trading has improved liquidity, cut transaction costs, and made market prices more efficient. They viewed the concerns about fraud as a public relations problem, not a real risk. The risks from programming errors, although real, are smaller than those from rogue human traders, some of whom have cost their employers billions of dollars in a few hours.

A recent report from an analyst at the Federal Reserve Bank of Chicago (Clark, 2012) urged traders to tighten up risk management procedures, but the procedures suggested were not new. Human programmers set limits and trading programs are cut off when they exceed those limits. The exchanges can institute "circuit breakers" which cut off trading when parameters are exceeded. These risk management procedures are imposed by both the trading companies and the exchanges. The problem is that applying these limits can slow trading even by fractions of second, so trading firms have an incentive to cut corners. Exchanges fear that imposing rigorous controls would cause them to lose business to competitors. The culture of the trading industry praises risk taking; risk managers have low status and are deprecated. But there are similar problems when humans conduct trading directly, indeed the psychology of irrational exuberance and panic may make things worse with humans at the helm. With human traders, there is also the psychological phenomenon of "normalization of deviance" where groups of people learn to tolerate unusual events such as spikes in markets until a really massive disaster occurs. Computers are less susceptible to these subtle social pressures.

At present, risk control parameters are largely set by human analysts, not by automated systems. There is an advantage, however, to making the process quicker and more complex, which favors having software set the limits. Trading software will evolve more and more sophisticated ways of monitoring the environment and incorporating the information into trading decisions, just as heart pacemakers, to be examined next, will evolve more and more complex ways of monitoring bodily processes. In the case of trading, the risks involved are only financial, they do not kill anybody, and the benefits seem to outweigh the costs, at least in the view of those most directly involved.

The data available for economic analysis are much better than they were a few decades ago, vastly beyond anything the Soviet planners or Salvador Allende's Chileans had. Economic modeling has also progressed, and its limitations are perhaps better understood. The challenge here is not to gather more economic data, but to integrate it with other kinds of data and use it to accurately predict and shape market trends and set government policies. Stronger AI programs are clearly needed here, and they are being funded by finance capitalists seeking to maximize market returns. Government agencies will need stronger and stronger AI to manage this economy.

4.4 PACEMAKERS

The first experimental heart pacemaker, invented in 1932, was an external mechanical device powered by a hand crank, with wires that attached to the body (Haddad and Serdijn, 2009). It could be set to 30, 60, or 120 beats per minute, and ran for 6 minutes before it needed to be cranked again. It served its purpose of demonstrating that electrical pulses could stabilize the human heart beat, but it was not until 1959 that the electronics were available to create an implantable pacemaker which beat a steady 60 beats per minute. The batteries lasted 5 years.

Since that time, pacemakers have rapidly become more complex because of the need to provide a flexible heartbeat that coordinates with the needs of the body's other systems. Modern pacemakers also synchronize ventricular stimuli to atrial activation. Although common sense would suggest that a simple mechanism would be more reliable than a complex one, this has not proven to be the case with pacemakers. Reliability has increased as the devices became more complex, due to continual improvement in the design and quality control of integrated circuits (Maisel, 2006). Research is also done to predict when pacemakers are likely to fail so they can be replaced well in advance.

Further developments in this field are not focused primarily on improving reliability because this is already so high. The main challenge in the field is to improve the integration with bodily processes such as blood pressure, blood glucose, and T-wave alternans (Brown University, n.d.). As implantable devices are developed to deal with other problems, such as insulin dispensers and defibrillators, the pacemakers will need to inter-act with these devices. This integration of multiple devices is a step-by-step progress toward a cyborgian future, when more and more bodily organs will be replaceable. This is moving rapidly in certain areas such as the development of artificial arms for injured American war veterans that respond to signals from the brain (Dao, 2012). Based on the experience with pacemakers, it seems that these replaceable body parts are likely to be quite reliable, especially those that depend primarily on integrated circuits. Simpler implants, such as hip or knee replacements, may have more reliability problems because they carry more physical stress. These artificial organs do not yet have the self-repairing capabilities of natural body parts, but they can be more readily replaced when they wear out or are damaged.

Pacemakers are very useful with the level of AI they now contain. But further development will require taking the information they collect and using it to predict oncoming heart disease and initiate treatments. Sensors will be needed in other parts of the body, and very sophisticated AI will be needed to analyze the data streams and make recommendations. This software will have to develop expertise going beyond that of any human specialist, because humans cannot analyze such massive continuous streams of data. It would have to learn for itself. At times, it would have to be trusted to act quickly in the event of heart attacks or other emergencies when there is no time to wait for human intervention. Of course, this raises challenges for security and reliability. These have already been explored in an episode of the very popular television series (Homeland, 2012) when terrorists assassinated the vice president of the United States by obtaining the password to the software controlling his pacemaker.

4.5 SELF-DRIVING MOTORCARS

Self-driving motorcars present an obvious risk of injury and death. But this risk also exists with human drivers and one of the rationales for developing self-driving automobiles is to cut the accident rate. Although the database is still small, even the early experimental technology has compared well for safety with human drivers (Thrun, 2006). Self-driving automobiles build on gradually evolving technologies including self-starters, automatic chokes, automatic transmissions, power steering, lane monitoring, automatic cruise control, blind spot monitoring, radar, and global positioning systems. Premium-class automobiles currently sold to the public carry 100 million lines of computer code, more than Boeing's new 787 Dreamliner (Vanderbilt, 2010).

Self-driving automobiles being tested by Google and others essentially bundle together these technologies. Their operation is then monitored with sophisticated algorithms including sequential Monte Carlo methods and Bayesian models in which latent variables are linked in Markov chains. A big challenge is to enable these methods to work quickly enough with the limited computing resources that can be fitted to a motorcar. These innovations are not dependent on any new theoretical breakthrough; actually the theory seems to emerge from the practice as the software generates more abstract models to facilitate its analysis. Costs are expected to come down significantly, especially as improved video camera technology will replace radar. Cars may be on the market in 5 years or so with limited self-driving capabilities, such as driving themselves during stop-and-go traffic or on limited access freeways (Levitan, 2012). Although fully self-driving motorcars have the potential to displace people employed as drivers, most cars are already driven by their owners or passengers, and the benefits are quite appealing. Safety regulations will probably keep human drivers on the job as backups for a long time.

4.6 SURVEILLANCE AND SOUSVEILLANCE

Although self-driving motorcars are generally viewed as desirable, surveillance by security agencies presents strong risks to privacy and human rights and is widely opposed (Ball and Webster, 2003; Brin, 1999; Goldman et al., 2009). These risks were anticipated in science fiction works such as *1984*. But the fiction writers did not usually explore the difficulties in processing and analyzing vast quantities of surveillance data, perhaps because the computerized data-mining techniques required did not exist until the computer age. Even today, security agencies find it much easier to monitor written communication than voice or video, although improvements in audio transcription and facial recognition may soon start to close that gap.

There is no realistic possibility of stopping technical advance in this field, making legal and political resistance by civil liberties and human rights groups essential. This has value in countries where the rule of law is observed. It may be very difficult to keep security agencies from using the technologies to collect information, but they may be restricted in using the information so gathered in legal proceedings.

Another approach is to try to avoid being observed or having communications intercepted, or to encrypt communications securely (Leistert, 2012). These methods probably cannot work if the authorities are willing and able to expend enough resources to defeat them in a given case. Current technology may not allow the authorities to monitor all communication fully, but advances in AI will continue to strengthen their capabilities. Ultimately, the only way to avoid detection may be to give up electronic communication altogether and go into deep hiding, as Osama bin Laden did, relying on trusted human couriers carrying paper messages. And even that ultimately failed.

The viable technical alternative is to turn the tables on the authorities, using the same techniques to observe them. This is sometimes called sousveillance, or surveillance from below (Brin, 1999; Mann, 2005). Related terms are participatory sensing (Shilton, 2010), cop watching, and cyberglogging (Mann et al., 2006). Of course, the power inequalities remain, but the information gathered through these techniques can be part of a social movement to empower people against the state. This has been highly effective at certain times when cell phone videos are posted to YouTube, Facebook, and other social media in repressive countries, generating mass movements.

Wearable and implantable cameras and computers can aid in this process, although existing groups rely mostly on cell phones. Resistance movements face the same problem as the authorities: how to process the masses of data that these technologies generate. Participatory sensing classes train community members to analyze the data as well as collect it, but this is a lot of work for only an occasional dramatic payoff.

There is a risk that the community activists themselves may become oppressive like the neighborhood Committees for the Defense of the Revolution in Cuba. But this seems a manageable risk in free countries where groups have no coercive power of their own and where anyone can conduct sousveillance and post their findings. Legal and political activism to protect these rights is an essential part of the process.

4.7 AI IN MEDICINE

Stand-alone systems designed to diagnose illnesses and recommend treatments were first developed in the 1970s (Patel et al., 2009), an approach sometimes optimistically described as "diagnosis without doctors" (Mazoué, 1990) or the "Greek Oracle" method (Miller and Masarie, 1990). These systems gave impressive results in experimental tests, but they were viewed as tools that physicians might use if puzzled, not as alternatives to diagnosis and prescription by human doctors (Maslove, 2011). Routine diagnostic needs did not justify the time needed to feed information into the AI systems. Critics argued that the process of diagnosis was too closely integrated with the process of treatment to be handled by computers (Miller, 1990). Physicians often initiated treatments before reaching a definitive diagnosis and took a wide range of information about the patient and his or her circumstances into account. Much of this knowledge was tacit rather than explicit and very difficult to formalize for input into an AI system. Although it was thought that AI might eventually be developed to carry out this complex process, doing so would require access to a much wider range of information than could be provided to the available programs at the time.

By the mid-1990s, the consensus in the field was that, despite a quarter decade of impressive technical accomplishment, the impact on medical practice had been minimal. A comprehensive review at the time (Coiera, 1996) concluded that "the modern study of artificial intelligence in medicine (AIM) is 25 years old. Throughout this period, the field has attracted many of the best computer scientists, and their work represents a remarkable achievement. However, AIM has not been successful—if success is judged as making an impact on the practice of medicine" (p. 363). The problem was not the complexity of the diagnostic process or the limitations of the computers. Rather, as the reviewer observed (Coiera, 1996), "it is now generally recognized that, before AIM research can produce systems that have a significant impact, a substantial information infrastructure has to be in place" (p. 365).

In the past 15 years, tremendous resources have been devoted to developing medical information systems, including electronic medical records, computerized physician order entry systems, and decision support systems (Blumenthal and Glaser, 2007; Harrison et al., 2007; Koppel, 2005). The task seems straightforward enough. Records that were previously kept on paper, including physicians' observations and analyses, need to be digitized using standardized categories and measurements. Information needs to be accurate and made available to medical practitioners in a timely manner. Therapeutic interventions and outcomes need to be continuously monitored and the outcome data made available to both clinicians and researchers in a timely and accurate fashion.

Implementing these systems has proven surprisingly difficult and has led to a multitude of unintended consequences (Rahimi et al., 2009). The difficulties fall into two main categories: entering and retrieving information, and integrating the information into patient care (Ash et al., 2004). Physicians and other practitioners often find the requirements for increased documentation burdensome, and resort to pasting in standardized boilerplate. Nurses find workarounds to avoid the burden of compliance with information systems, such as taking bar code tags to a central nursing station and clicking on them there instead of at each patient's bedside when services are provided. AI systems generate too many warnings of possible but highly unlikely drug interactions, leading physicians to generally ignore them. Informal interactions and redundant checks are minimized, increasing efficiency at the cost of a greater risk of serious errors. Errors creep in because of different use of abbreviations or measurement conventions. Expenses go up rather than down because electronic systems make it easier to submit bills for multiple procedures.

In the United States, these problems may be complicated by the multiplicity of insurance and billing systems, although billing procedures were computerized early. But the difficulties have been very great in the United Kingdom (Fraser and Stelzer, 2009) despite the single payer public system. The fundamental problem is integrating computer technology with a complex social system that depends heavily on informal cultural practices and tacit knowledge. Some of these problems can be resolved with technology such as smart wearable systems that automatically monitor body signs and input the information into the information systems (Chan, 2012). Indeed, the information so gathered may be superior in many ways to that collected by doctors and nurses. But humans will not be taken out of the loop for the foreseeable future, and successfully integrating AI into the system requires a sociotechnical approach that focuses on how technology is actually used in human social systems.

Advances in medical informatics have also raised new privacy issues, as information is gathered from sensors attached to the body or collected with mobile devices and shared easily with providers, insurers, employers, and patients (Malin et al., 2013). The ready availability of information has great potential for improving health care, but it also sometimes leads patients to lose trust in their providers and forego treatment or seek help outside the system, especially for mental health issues and other stigmatized conditions. Many of the issues are ethical and legal, as well as technical, and involve conflicting interests that are difficult and time consuming to resolve.

As smart wearable systems become more widely used, sophisticated AI will be needed to analyze and act on the information they provide. Health informatics will create vast new streams of data about the effectiveness of treatments and about costs and inefficiencies in the system. Writing in 1990, Randolph Miller (1990) responded to James Mazoué's (1990) "diagnosis without doctors" proposal by arguing that "for the remainder of this century and at the beginning of the next, humans will be the primary purveyors of medical diagnosis" (p. 582). He argued that diagnosis was an "art" rather than a precise science because it involved many steps including gathering data, integrating it into plausible scenarios and refining hypotheses, initiating therapy, monitoring the results, and evaluating the effects. In 1990, digital records of all these steps were not kept, nor were the AI programs of the time prepared to use it. We are rapidly approaching the time when the data will be accessible, creating great new opportunities to develop the stronger AI systems needed to make use of it.

4.8 CONCLUSIONS

The cases we have examined are quite different, and the risks, benefits, and technical challenges they present vary widely. Setting up a giant computer system to run an entire national economy was long ago abandoned for technical reasons, and almost no one considers reviving it today. Just because a project might be technically feasible does not mean that it will be implemented. Human institutions are still firmly in charge of that. With computerized financial trading, the primary challenge is analytical. Tremendous quantities of data are already available; the challenge is to come up with algorithms that can actually beat the market over long periods of time. In medicine, the primary challenge is building the information system to feed data into algorithms that are already quite sophisticated. With surveillance, the primary challenges are political, finding ways to gather information that is really needed to fight terrorism and solve crime without compromising civil liberties. Self-driving motorcars, however, are almost entirely a technical challenge without raising political issues.

In all of the areas we have examined, it seems clear that the next 10–25 years will be characterized by a merger of narrow AI and strong AI approaches as developers develop intelligent agents that can interact actively and creatively with complex environments. These agents will not be the humanlike robots of science fiction, at least not for several decades. At some point, it may become possible to create a robot that could drive itself to work, engage in electronic trading at a bank in the morning, diagnose and treat patients in a hospital in the afternoon, volunteer for a cyberglogging collective in the evening, then fight terrorists at night. But there is no convincing reason to concentrate all these skills in one electronic individual. Rather, resources will be used to gradually strengthen the capabilities of AIs designed for narrow tasks, gradually linking them together to form a more general AI.

Risk management will also follow this approach. Programs will be developed to manage specific risks: sousveillance to counter surveillance, automated traffic cops to police self-driving cars, automated risk-limiting computers at stock markets, and sensors to monitor bodily functioning and detect when pacemakers and other implants malfunction. There will certainly be failures and risks, but they will be less catastrophic than they might be if a single all-powerful AI were developed.

If this vision of the development of AGI is correct, we will not have one global brain on one mainframe computer, but millions of AIs with specialized but increasingly complex functionality. More general intelligence will emerge as these specialized AIs are linked together in complex and varied ways. The human-level AGI that eventually develops may be similar to the Siri program on the iPhone but much smarter, interfacing with dozens or hundreds of AGI experts, as well as with human beings, in carrying out its tasks. By working with other agents, such an assistant could awaken its owner, deliver the morning's news and messages, order the kitchen to prepare a nutritionally appropriate breakfast, select clothing to fit the day's weather and the scheduled events, order the car to retrieve itself from valet parking, monitor the owner's vital signs, and interface with the computers at the owner's workplace. In doing these things, it could choose to interact with a large number of collaborating AIs, selecting those with high reliability as well as advanced functionality, as judged by automated crowd sourcing techniques.

All of these AIs will be constantly interacting with each other, seeking improved partners, and maximizing their functionality and reliability. Each step in this development process will involve new risks and strategies for containing them will evolve along with the new capabilities. The AGI that will emerge from this process will rapidly exceed human levels in performing many tasks, although it will have no reason to evolve the exact set of abilities—and vulnerabilities—found in human beings.

The small-N comparative method has some value for projecting trends a decade or two into the future. The options for the more distant future are best addressed with philosophical and speculative methods. However, longer term futurist analyses should take into account the way things are likely to develop in the more foreseeable future. It seems certain that more generally intelligent AI agents will continue to develop for a period of years before the much feared development of a superhuman AGI with the capability of modifying itself into new forms and presenting an existential risk to the human species. When (or if) this superhuman AGI develops, it will incorporate the capabilities and risk management systems of earlier AGI agent systems, just as the human brain evolved from the brains of more primitive ancestors.

REFERENCES

Abbott, A. (2004). *Methods of Discovery: Heuristics for Social Science*. New York: Norton.

Ash, J., Berg, M., and Coiera, C. (2004). Some unintended consequences of information technology in health care: The nature of patient care information system-related errors. *Journal of the American Medical Informatics Association*, 11, 104–112. doi:10.1197/jamia.M1471.

Ball, K. and Webster, F. (2003). *The Intensification of Surveillance: Crime, Terrorism and Warfare in the Information Era*. London: Pluto Press.

Baradit, J. (2008). *SYNCO*. Santiago, Chile: Ediciones B.

Blumenthal, D. and Glaser, J. (2007). Information technology comes to medicine. *New England Journal of Medicine*, 356, 2527–2534. doi:10.1056/NEJMhpr066212.

Brin, D. (1999). *The Transparent Society*. New York: Basic Books.

Brown University, Division of Biology and Medicine. (n.d.). "The Near Future of Pacemaker Technology," and "The Distant Future of Pacemaker Technology." Retrieved from http://biomed.brown.edu/Courses/BI108/BI108_1999_Groups/ Cardiapacing_Team/nearfuture.html and http://biomed.brown.edu/Courses/ BI108/BI108_1999_Groups/Cardiapacing_Team/distantfuture.html.

Chan, M., Estève, D., Fourniols, J. Y., Escriba, C., and Campo, E. (2012). Smart wearable systems: Current status and future challenges. *Artificial Intelligence in Medicine*, 56(3), 137–156. doi:10.1016/j.artmed.2012.09.003.

Clark, C. (2012). How to keep markets safe in the era of high-speed trading. *Chicago Fed Letter*, October. Retrieved from http://www.chicagofed.org/ webpages/publications/chicago_fed_letter/2012/october_303.cfm.

Coiera, E. (1996). Artificial intelligence in medicine: The challenges ahead. *Journal of the American Medical Informatics Association*, 3(6), 363–366. Retrieved from http://171.67.114.118/content/3/6/363.full.pdf.

Dao, J. (2012). Learning to accept, and master, a $110,000 mechanical arm. *New York Times*, November 26. Retrieved from http://www.nytimes.com/2012/11/27/us/ prosthetic-arms-a-complex-test-for-amputees.html?pagewanted=all&_r=1&.

Dieterich, H. (2007). *Hugo Chávez y el Socialismo del Siglo XXI*. Edición Digital. Retrieved from http://www.rebelion.org/docs/55395.pdf.

Foresight: The Future of Computer Trading in Financial Markets. (2012). Final Project Report. London: The Government Office for Science. Retrieved from http://www.bis.gov.uk/assets/foresight/docs/computer-trading/12-1086- future-of-computer-trading-in-financial-markets-report.pdf.

Fraser, N. and Stelzer, I. (2009). The British Fiasco. *The Weekly Standard*, July 27. Retrieved from http://www.weeklystandard.com.

Gerovitch, S. (2002). *From Newspeak to Cyberspeak*. Cambridge, MA: MIT Press.

Goertzel, T. (2009). Rethinking socialism in Today's Latin America. In Font, M., ed., *A Changing Cuba in a Changing World* (pp. 287–294, Chapter 16). New York: Bildner Center for Western Hemisphere Studies. Retrieved from http://web. gc.cuny.edu/bildnercenter/publications/documents/Goertzel16_000.pdf.

Goldman, J., Shilton, K., Burke, J., Estrin, D., Hansen, M., Ramanathan, N., Reddy, S., Samanta, V., Srivastava, M., and West, R. (2009, May). *Participatory Sensing: A Citizen-Powered Approach to Illuminating Patterns That Shape Our World*. Woodrow Wilson International Center for Scholars White Paper. Retrieved from http://www.wilsoncenter.org/sites/default/files/participatory_sensing.pdf.

Guzmán, J. (2003). Proyecto Synco: El Sueño Cibernético de Salvador Allende. *Clinic*, July 10. Retrieved from http://www.theclinic.cl/2011/08/28/el-sueno-cibernetico-de-allende/.

Haddad, S. A. and Serdijn, W. (2009). *Ultra-Low Power Biomedical Signal Processing: An Analog Wavelet Filter Approach for Pacemakers*. Dordrecht, the Netherlands: Springer.

Harrison, M., Koppel, R., and Bar-Lev, S. (2007). Unintended consequences of information technologies in health care—An interactive sociotechnical analysis. *Journal of the American Medical Informatics Association*, 14(5), 542–549. doi:10.1197/jamia. M2384.

Homeland. Television Series. (2012). Episode 12. Retrieved from http://www.sho. com/sho/homeland/season/2/episode/12.

Koppel, R., Metlay, J. P., Cohen, A., Abaluck, B., Localio, A. R., Kimmel, S. E., and Strom, B. L. (2005). Role of computerized physician order entry systems in facilitating medication errors. *Journal of the American Medical Association*, 293(10), 1197–1203. doi:10.1001/jama.293.10.1197.

Kurzweil, R. (2005). *The Singularity Is Near*. New York: Viking Press.

Lee, D. (2012). David Cameron testing app to aid government decisions. *BBC Technology News (Online)*, November 7. Retrieved from http://www.bbc. co.uk/news/technology-20240874.

Leistert, O. (2012). Resistance against cyber-surveillance within social movements and how surveillance adapts. *Surveillance and Society*, 9(4), 441–456. Retrieved from http://www.surveillance-and-society.org/.

Levitan, D. (2012). Self-driving cars: Coming soon to a highway near you. *Yale Environment 360*, July 23. Retrieved from http://e360.yale.edu/feature/self-driving_cars_coming_soon_to_a_highway_near_you/2554/.

Maisel, W. (2006). Pacemaker and ICD generator reliability: Meta-analysis of device registries. *Journal of the American Medical Association*, 295(16), 1929–1934. Retrieved from http://www.ncbi.nlm.nih.gov/pubmed/16639052.

Malin, B., Emam, K., and O'Keefe, C. (2013). Biomedical data privacy: Problems, perspectives, and recent advances. *Journal of the American Medical Informatics Association*, 20(1), 2–6. doi:10.1136/amiajnl-2012-001509.

Mann, S. (2005). Sousveillance and cyborglogs: A 30 year empirical voyage through ethical, legal and policy issues. *Presence: Teleoperators and Virtual Environments*, 14(6), 625–646. Retrieved from http://www.mitpressjournals. org/doi/abs/10.1162/105474605775196571.

Mann, S., Fung, J., and Lo, R. (2006). Cyborglogging with Camera Phones: Steps toward Equiveillance. In *Proceedings of the ACM Multimedia*. Santa Barbara, CA: ACM. Retrieved from: http://www.eyetap.org/papers/docs/art0905s-mann.pdf.

Markoff, J. (2012). Scientists see promise in deep-learning programs. *New York Times*, November 23. Retrieved from http://www.nytimes.com/2012/11/24/science/scientists-see-advances-in-deep-learning-a-part-of-artificial-intelligence.html.

Maslove, D. (2011). Computerized physician order entry in the critical care environment: A review of current literature. *Journal of Intensive Care Medicine*, 26(3), 165–171. doi:10.1177/0885066610387984.

Mazoué, J. (1990). Diagnosis without Doctors. *Journal of Medicine and Philosophy*, 15(6), 559–579. Retrieved from http://www.ncbi.nlm.nih.gov/pubmed/2290071.

Medina, E. (2011). *Cybernetic Revolutionaries: Technology and Politics in Allende's Chile*. Cambridge, MA: MIT Press.

Miller, R. (1990). Why the standard view is standard: People, not machines, understand patients' problems. *The Journal of Medicine and Philosophy*, 37, 581–591. Retrieved from http://www.ncbi.nlm.nih.gov/pubmed/2290072.

Miller, R. A. and Masarie, F. E. (1990). The demise of the "Greek Oracle" model for medical diagnostic systems. *Methods of Information in Medicine*, 29(1), 1–2. Retrieved from http://www.ncbi.nlm.nih.gov/pubmed/2407929.

Patel, V., Shortliffe, E., Stefanelli, M., Szolovits, P., Berthold, M., Bellazzi, R., and Abu-Hanna, A. (2009). The coming age of artificial intelligence in medicine. *Artificial Intelligence in Medicine*, 46(1), 5–17. doi:10.1016/j.artmed.2008.07.017.

Rahimi, B., Vimarlund, V., and Timpka, T. (2009). Health information system implementation: A qualitative meta-analysis. *Journal of Medical Systems*, 33(5), 359–368. Retrieved from http://link.springer.com/article/10.1007%2Fs10916-008-9198-9.

Rampell, R. (2013). Raging (again) against the robots. *New York Times*, February 2. Retrieved from http://www.nytimes.com/2013/02/03/sunday-review/raging-again-against-the-robots.html?_r=0.

Shilton, K. (2010). Participatory sensing: Building empowering surveillance. *Surveillance and Society*, 8(2), 131–150. Retrieved from http://www.surveillance-and-society.org/ojs/index.php/journal/issue/view/Empowerment.

Thrun, S., Montemarlo, M., Dahlkamp, H., Stavens, D., Aron, A., Diebel, D., ..., Stang, P. (2006). Stanley: The robot that won the DARPA grand challenge. *Journal of Field Robotics*, 23, 661–692. Retrieved from http://www-robotics.usc.edu/~maja/teaching/cs584/papers/thrun-stanley05.pdf.

Vanderbilt, T. (2010). Let the robot drive. *Wired Magazine*. Retrieved from http://www.wired.com/magazine/2012/01/ff_autonomouscars/all/.

Wikipedia. (2013). *Strong-AI*. Retrieved from http://en.wikipedia.org/wiki/Strong_ai.

Wiseman, P., Condon, B., and Fahey, J. (2013). Practically human: Can smart machines do your job? *Associated Press: The Big Story*, January 24. Retrieved from http://bigstory.ap.org/article/practically-human-can-smart-machines-do-your-job-1.

CHAPTER 5

Limitations and Risks of Machine Ethics

Miles Brundage

CONTENTS

ABSTRACT Many authors have proposed constraining the behavior of intelligent systems with "machine ethics" to ensure positive social outcomes from the development of such systems. This chapter critically analyzes the prospects for machine ethics, identifying several inherent limitations. Although machine ethics may increase the probability of ethical behavior in some situations, it cannot guarantee it due to the nature of ethics, the computational limitations of computational agents, and the complexity of the world. Additionally, machine ethics, even if it were to be "solved" at a technical level, would be insufficient to ensure positive social outcomes from intelligent systems.

5.1 INTRODUCTION

In response to the increasing sophistication and social impact of artificial intelligence (AI)-based technologies, some authors have proposed that software agents and robots be imbued with explicit ethical principles to govern their behavior (Wallach and Allen, 2010). This area of research, known as "machine ethics," incorporates insights from normative ethics, AI, and other disciplines and seeks to instantiate models of ethical reasoning in machines in order to ensure ethical machine behavior as well as to explore the nature and computability of human ethics. In addition, other authors concerned with the risks posed by future greater-than-human intelligence machines have characterized machine ethics as one among multiple strategies for ensuring positive social outcomes from the creation of such machines. Yudkowsky (2001), for example, argues that failing to develop a sufficiently well-formulated computational instantiation of morality will result in catastrophic outcomes from developing superintelligent machines, because the instrumental goals of self-improving systems seeking to maximize some arbitrary utility function will at some point come into conflict with human interests.

What these literatures (referred to here as "traditional machine ethics" on the one hand, but "artificial general intelligence (AGI) ethics" on the other, though the two have cross-pollinated to some extent) have in common is the apparent belief that the ethical behavior "problem" can be "solved" to some extent through philosophical and technological innovation. Although this chapter does not seek to undermine the desirability of such a solution, in principle, it points to a number of reasons to believe that such a project will necessarily fail to guarantee ethical behavior of a given AI system across all possible domains. The intrinsic imperfectability of machine ethics has been suggested by several authors, such as Allen et al. (2000); this chapter seeks to synthesize, deepen, and extend these concerns, and draw out some of their implications for the project of machine ethics *vis-à-vis* the possible social outcomes from AGI. Although such inevitable imperfection may be acceptable to traditional machine ethicists, who have often explicitly acknowledged and accepted that machines, like humans, will inevitably make some mistakes, it presents a fundamental problem for the usefulness of machine ethics as a tool in the toolbox for ensuring positive outcomes from powerful computational agents (Sotala and Yampolskiy, 2013).

The chapter will proceed as follows: first, a diverse set of inherent limitations on the machine ethics project which come from the nature of ethics, the nature of computational agents in general, and the nature of the world will be identified. Next, some specific categories of machine ethics and AGI ethics proposals will be reviewed, finding that they are susceptible to the limitations identified in Section 5.2. A number of possible failure modes for machine ethics will be summarized, that is, ways in which instantiating ethical principles in a powerful AGI system could result in outcomes that are unintended and possibly even catastrophic. Finally, machine ethics will be characterized as an insufficient (albeit potentially helpful) tool for ensuring positive social outcomes from AGI, because other factors (such as cybersecurity, human decisions, and systemic risks) will also have to be reckoned with in the creation of human-level or greater-than-human intelligence machines.

5.2 LIMITATIONS FROM THE NATURE OF ETHICS

This section will survey some of the literature in normative ethics and moral psychology, which suggest that morality does not lend itself to an algorithmic solution. This is not to say that humans and machines cannot improve their moral behavior in many situations by following the prescriptions of something akin to an algorithm—indeed, there is widespread agreement on at least some moral rules (Gert, 2007). However, these rules are often ambiguous and should sometimes be broken, and there is persistent disagreement about the conditions in which such exceptions should be made, as well as broad agreement that some specific ethical domains are still problematic despite the best efforts of philosophers. Importantly, for the present discussion, this "unsolved" nature of ethics may not be a transient condition owing to insufficient rational analysis, but rather a reflection of the fact that the intuitions on which our ethical theories are based are unsystematic at their core, which creates difficulties for the feasibility of machine ethics.

Researchers dating back at least to Darwin in *The Descent of Man* (1871) have attempted to explain human moral judgments as deriving, at least in part, from evolutionary processes. Recently, Boehm (2012) has distilled much anthropological and evolutionary psychological evidence to argue that many of our intuitions about morality have their origins in prehistoric humans' and prehuman primates' attempts to suppress antisocial behavior. Under this reading of the scientific evidence, virtue, altruism, shame, and other morally salient concepts and emotions exist because those who are alive today descended from humans that adapted, genetically and culturally, to the demands of social cooperation. Findings in the social psychology and neuroscience of morality bear out the deep connection between ethical judgments and evolutionary requirements.

Cushman et al. (2010) articulate an emerging consensus in moral psychology work that much of our moral judgments can be conceptualized through a "dual-process" model. Under this framework, humans rely to varying extents, depending on the situation at hand and factors such as cognitive load, on an intuitive moral system and a more deliberate moral system. Moreover, this dual-process model of moral psychological processes roughly maps onto two major camps in moral philosophy: deontological (means-based) and consequentialist (ends-based) decision making. Given these distinct ways by which humans arrive at moral judgments, and the fact that ethics is, at least in part, a project aimed at systematizing moral judgments into a rational framework, dual-process moral psychology is of great importance to machine ethicists. These findings are recent and have been contested on grounds ranging from their empirical bases (Klein, 2011) to their normative implications (Berker, 2009), but at the moment, they seem to offer a plausible explanation for the persistence of ethical quandaries and the difficulty of articulating an exceptionless and well-specified moral theory.

Having assessed some empirically based reasons to suspect that human moral judgments may not lend themselves to a consistent computational specification, the remainder of this section will consider barriers to reliable machine ethics that are discussed in the literature on ethics proper without commenting on the underlying moral psychology involved in these issues. Comprehensive moral theories that have been put forward so far are inadequate in varying ways and to varying extents. An exhaustive survey of the state of the debate in normative ethics is beyond the scope of this chapter, but a few brief points on the two leading branches of ethical theories (deontological and consequentialist) will illuminate some of the challenges facing a hypothetical machine ethicist seeking to instantiate an "off-the-shelf" ethical theory in a machine (issues facing "bottom-up" approaches based on, e.g., case-based reasoning and machine learning will be discussed later).

Consequentialist theories of ethics have been criticized as inadequate for a variety of reasons (Pojman, 2005). These include the claim that they cannot sufficiently account for the moral significance of an individual's personal social commitments (e.g., to friends and family) and life projects; that they impose excessive demands on individuals to contribute to others' welfare (Wolf, 1982); that it seems to arrive at unacceptable conclusions in certain situations; and that it fails to give sufficient consideration to the separateness of persons, distributional justice considerations, and individual rights (Williams, 1973). Some of these objections have been leveled at consequentialism in general and others at specific variations thereof; additionally, a wide variety of consequentialist theories have been developed to attempt to grapple with the concerns that have been raised, but none have achieved anything remotely resembling a consensus in philosophy.

Deontological theories of ethics also have their limitations. Some notable critiques include the argument that deontological theories give the wrong answers in situations involving extreme trade-offs between the interests of the few and the interests of the many, and can thus produce catastrophic moral results; that deontological theories cannot adequately resolve conflicts between duties; and that deontology collapses into consequentialism because an actor who opposes producing harm X is rationally committed to reducing the amount of X in the world.

Each of the theories thus described has articulate defenders, as do other ones that attempt to synthesize the insights of each (Parfit, 2011). However, for the purposes of this discussion, I seek merely to highlight that a machine ethicist seeking for input from normative ethicists will receive a wide range of (often heavily qualified) responses, and to the knowledge of the author, each proposed solution is likely vulnerable to at least one of the objections above. This pervasive disagreement about and unease with comprehensive moral theories has been noted by many authors (Gert, 2007)—indeed, even the defenders of specific theories mentioned above typically note the difficulty of their respective frameworks in addressing certain issues and the need for further analysis. An additional overarching issue facing comprehensive moral theories is ambiguity. Although moral theories have varying degrees of specificity in terms of both their criteria of moral rightness and proposed decision procedures, they must be augmented by knowledge of and sensitivity to the relevant domains in which an individual finds himself or herself. Ethics on the battlefield are very different from ethics in the shopping mall—thus, although applied ethical decisions can be loosely inspired by overarching moral theories, much work remains to be done in order to flesh out what, for example, pleasure, welfare, beneficence, or nonmalevolence actually mean in a particular domain, and this derivation and specification introduces new challenges (some of which will be discussed in Section 5.3).

Another challenge to ethical systematization is posed by the plurality of values which arguably should (and empirically do) motivate people. This issue is related to, but can also be distinguished from, the objections to comprehensive moral theories summarized above. Berlin notes the cultural and historical variation in the values that people do, in fact, hold (Berlin, 1990), and argues that the move to encompass all of what ought to be done under the umbrella of a single value has historically been associated with oppression and violence. Ross (1988), whose ethical views (drawn on by some machine ethicists) do not fit cleanly into either the deontological or consequentialist categories, argues based on an analysis of ethical intuitions that multiple *prima facie* duties exist, none of which is reducible to the others. The challenge of value pluralism within comprehensive moral theories has also been noted: for example, Shaw (1999) surveys the literature on utilitarianism (a variant of consequentialism) and finds no clear answer to the question of what it is that ought to be maximized by a utilitarian (that is, pleasure, welfare, the satisfaction of preferences, and so on). Shulman et al. (2009) arrive at a similar conclusion. Haidt (2012) argues based on consequentialist grounds that in the public domain, people should invoke a plurality of moral "foundations," some oriented more toward individual autonomy and others more oriented toward group cohesion and welfare, because each on its own adds something important to normative discourse. Similar arguments have been made for thousands of years, in both secular and religious traditions (see, e.g., the Ten Commandments, Aristotle's virtues, etc.) and incorporating these insights into a machine ethics system would seem to entail unpredictable behavior or paralysis in the case of value trade-offs.

Next, regardless of whether a comprehensive moral theory is sufficient to capture the "right" answers to a wide range of ethical, there remain some unresolved issues in ethics that must be grappled with if ethical machines are to be deployed in a flexible way across many domains. These issues inspire many of the counterexamples often used against comprehensive moral theories, and are also investigated in and of themselves by ethicists, without clear resolution to date. Crouch (2012) highlights several of these "unsolved problems" in ethics, including population ethics, small probabilities of huge amounts of value, issues associated with the possibility of infinite value, moral uncertainty, the relationship between intuitions and theoretical virtues, prevention of wrongs versus alleviating suffering, and the assignment of moral value to various entities. Although many philosophers have preferred answers to these questions, these issues remain deeply controversial and a machine making decisions in an open, complex environment over the long term will need to grapple with them and may arrive at conclusions that, based on the present understanding of ethicists, will be vulnerable to compelling objections.

Finally, the (possible) existence of genuine moral dilemmas is problematic for the machine ethicist. The theoretical and actual possibility of genuine moral dilemmas is a matter of disagreement among ethicists (Gowans, 1987). Some argue for the existence of such dilemmas based on some of the difficulties discussed earlier (such as value pluralism and conflicts within and between ethical theories—see, e.g., Sinnott-Armstrong, 1988) and on an argument for rational regret. Williams (1973) argues that even when one is confident that one has made the right ethical decision, doing so often requires making trade-offs between fundamentally distinct considerations, leaving an ethical "remainder" that can rationally be regretted because it was a different sort of thing than that which was ultimately prioritized. By contrast, others argue based on the principles of deontic logic and other considerations that moral dilemmas are not logically possible within a rational theory of ethics. Yet whether such dilemmas can be considered "genuine" or merely perceived as such from an internal or external perspective, conflicts between duties, values, and interests are commonplace in real-life situations that ethical machines will find themselves in, and as such will pose difficult challenges for creating systems that will be *perceived as* having acted morally after the fact.

5.3 LIMITATIONS ARISING FROM BOUNDED AGENTS AND COMPLEX ENVIRONMENTS

Given an arbitrary ethical theory or utility function, a machine (like a human) will be limited in its ability to act successfully based on it in complex environments. This is in part due to the fact that ethical decision making requires an agent to estimate the wider consequences (or logical implications) of his or her actions, and possess relevant knowledge of the situation at hand. Yet humans and machines are limited in their perceptual and computational abilities, and often only have some of the potentially relevant information for a given decision, and these limitations will create possible failure modes of machine ethics.

Computational and knowledge limitations apply to ethics in different ways depending on the ethical theory involved. Consequentialist theories, for example, are quite explicitly dependent on the ability to know how one's actions will affect others, though there is still much heterogeneity here, such as the distinction between objective utilitarianism (which prescribes acting in a way that, in fact, maximizes good outcomes) and subjective utilitarianism (which emphasizes the expected outcomes of one's actions). Deontological theories, by contrast, are not explicitly about foreseeing the outcomes, but computational limitations are related to deontology in at least two ways. First, to even know in the first place that a given action is, for example, consistent with a given deontological duty may require some knowledge and analysis of the situation and actors involved, not all of which will necessarily be apparent to the actor. Second, some deontological theories and rule consequentialist theories (which prescribe acting on the set of rules that, if universally accepted or adopted, would lead to the best outcomes) require judgments about the logical implications of a given decision if it were to be performed by many or all actors, that is, the universalizability of a moral judgment. As such, the knowledge and information processing limitations of an artificial agent will be relevant (though perhaps to varying degrees) regardless of the specific ethical theory invoked.

There is wide recognition in AI, cognitive science, and other fields that humans and machines are not able to act in a fully rational manner in nontrivial domains, where this is defined as maximizing a given utility function by processing information from the information and acting in one's environment based on it, although there are efforts to move closer and closer to approximating such rationality. This inherent limitation is due to the potentially infinite courses of action available to an agent, and the consequent inability to exhaustively analyze all of them; the inability to obtain and store all potentially relevant facts; and the intrinsically uncertain relationship between chosen actions and consequences when the environment has complex dynamics. Consequently, some have suggested that concepts such as bounded rationality (acting more or less rationally given one's computational limitations) and satisficing (acting in a way that leads to satisfactory or "good enough" outcomes) are better models for thinking about decision making by agents such as humans and machines. Indeed, Wang (2006) argues that the ability to act on the basis of insufficient knowledge and resources is the essence of intelligence. Gigerenzer (2010) and others note that heuristics, while deviating from normative theories of rationality, in fact work quite well in many situations in which we find ourselves, that is, they have adaptive value. The literature on human cognitive heuristics may seem removed from machine ethics at first glance—after all, as Yudkowsky (2007) and others note, the goal of AI is not merely to replicate human behavioral patterns but to improve upon them. But this is not the case. Insofar as humans serve as the "proof of concept" for intelligence and some AI theorists seek to develop cognitive architectures that are inspired by human cognition, we may ultimately see ethical machines that exhibit similar cognitive heuristics to humans. Additionally, because machines will learn and evolve in similar environments (namely, complex social ones) to those in which humans evolved, even if AI does not explicitly seek to model human behavior, some of the same heuristics that benefitted humans may ultimately creep in. Regardless, the points made here do not depend on the existence of any particular heuristic being instantiated in a machine, but rather it suffices to note that machines are and will remain limited in their possession and processing of knowledge.

Computational and knowledge limitations enter into ethical decision making in several ways. First, one can think of the drawing up of a list of possible actions as a search problem. Under this framing, it is impossible to exhaustively search the space of all possible actions in any nontrivial domain that could lead to better and better outcomes with regard to a specific ethical theory, though this does not preclude responding to binary ethical decisions. Heuristics will be required to select a set of actions to evaluate, and although these may be useful in many situations (as they are for humans), important options for navigating moral dilemmas may consequently be overlooked (Gigerenzer, 2010). Gigerenzer argues that satisficing is the norm in real-life ethical decision making. Second, given a particular problem presented to an agent, the material or logical implications must be computed, and this can be computationally intractable if the number of agents, the time horizon, or the actions being evaluated are too great in number (this limitation will be quantified later and discussed in more detail later in the section). Specifically, Reynolds (2005, p. 6) develops a simple model of the computation involved in evaluating the ethical implications of a set of actions, in which N is the number of agents, M is the number of actions available, and L is the time horizon. He concludes:

> It appears that consequentialists and deontologists have ethical strategies that are roughly equivalent, namely $O(MN^L)$. This is a "computationally hard" task that an agent with limited resources will have difficulty performing. It is of the complexity task of NP or more specifically EXPTIME. Furthermore, as the horizon for casual ramifications moves towards infinity the satisfaction function for both consequentialism and deontologism become intractable.

While looking infinitely to the future is an unreasonable expectation, this estimate suggests that even a much shorter time horizon would quickly become unfeasible for an evaluation of a set of agents on the order of magnitude of those in the real world, and as previously noted, an infinite number of actions are always available to an agent. This argument has also been made qualitatively in the works of Allen et al. (2000) and Wallach and Allen (2010). Goertzel has also outlined a similar argument to Reynolds's (2006).

Lenman (2000) has spelled out the implications of such limitations for consequentialism, arguing that a person is only aware of a small number of the possible consequences of a given decision, and that the longer term consequences may in fact dominate the ultimate ethical impact, making consequentialism impossible to fully adhere to—for example, a seemingly trivial decision made while driving that slightly affects traffic patterns could have implications for who ends up meeting whom, when children are conceived and their consequent genetic makeup, and so on, which are far more significant than the driver's immediate impacts, yet are utterly impossible to foresee and act upon. Third, an agent could lack knowledge that is of great ethical significance for the decision. This is related to the "frame problem" in AI and has been extended to the domain of ethics by Horgan and Timmons (2009), who note that small changes in the context of a given decision could have great ethical import yet cannot be captured in a single ethical principle.

Assuming that an agent had a well-specified utility function for ethics and exhibited normative rationality in dealing with ethical dilemmas, the agent would nevertheless run into problems in complex task domains. Social systems exhibit many of the characteristics of complex systems in general, such as nonlinear interactions, heavy-tail distributions of risks, systemic risks introduced by failure cascades in coupled systems, self-organization, chaotic effects, and unpredictability (Helbing, 2010). These are not avoidable features of an agent making assessments of the systems, but rather are fundamental to the complex world we inhabit. In addition, the adoption of technologies adds additional elements of complexity to social analysis, because technologies influence social outcomes at multiple scales that are not reducible to a simple predictive model (Allenby and Sarewitz, 2011). These complexities pose challenges for ethical decision making by making the system being analyzed beyond the knowledge and computational scope of any agent making the decision—that is, any assessment by an artificial agent, such as that buying stock in a particular company would be beneficial for its owner, or that diverting a train to save the lives of several people standing on a train track at the expense of one person, is necessarily no more than educated guesses that could be wrong in the face of "black swan" events that fly in the face of predictive models that made sense beforehand (Taleb, 2007). Additionally, an agent could be ignorant of the fact that he or she will have a nonlinear effect on a complex system—such as that selling a certain stock at a certain volume, combined with other trading activities, will induce a "tipping point" and runaway sales of that stock, with ethical ramifications dwarfing that of the originally intended sale. Lack of knowledge about the intentions of other agents, too, poses a challenge, but even given such knowledge, a superintelligent agent could not solve the predictability problem: for example, using a relatively simple model of a multiagent system (compared to the real ones we inhabit), researchers (Tosic and Agha, 2005) have found that

> counting the number of possible evolutions of a particular class of CFSMs [communicating finite state machine] is computationally intractable, even when those CFSMs are very severely restricted both in terms of an individual agent's behaviour (that is, the local update rules), and the inter-agent interaction pattern (that is, the underlying communication network topology).

Discussing such complexities, Allenby (2009, p. 2) notes that "CASs [complex adaptive systems] thus pose a challenge to the existing ethical approaches most familiar to scientists and technologists." Although these points suggest difficulties for any ethical agent, they particularly pose a challenge for an artificial agent seeking to act across multiple domains and to intervene at a large scale in complex systems—for example, by seeking to maximize some global utility function such as human welfare, rather than more limited action at a small scale.

Two final points can be made about the limitations of machine ethics *vis-à-vis* action in a social environment. First, as Müller notes (2012), to improve their effectiveness, artificial agents will need to learn and evolve over time, which requires more autonomy and, correspondingly, less human control of the agent in question. Likewise, Wallach and Allen (2010) and others have noted the importance of integrating top-down and bottom-up approaches to machine ethics, the latter requiring learning from specific cases and experiences. Representative training data and experiences of an artificial agent will thus be essential for the "successful" development of machine ethics. Yet it is impossible to expose an agent to an infinite array of possible ethical situations, and care will be needed to ensure that the lessons being learned are those intended by the designers of the system. Additionally, as the ethical principles appropriate to a given domain are often unique, it is unclear when and how one could determine that an ethical training process were ever "complete." Second, the complexity of artificial agents may pose challenges to humans' ability to understand, predict, and manage them. Already, there are technological systems that elude human understanding, and the development of human-level or greater-than-human intelligence systems would exacerbate this problem and pose difficulties for the allotment of trust to such systems.

5.4 CURRENT APPROACHES TO MACHINE/AGI ETHICS

This section will summarize some elements of the machine and AGI ethics literatures and note how current approaches are vulnerable to the limitations outlined above. Notably, few of these approaches have been suggested by researchers as the be-all-end-all for machine ethics—rather, the field is at an early stage, and these approaches are mostly being presented (by the researchers behind them, and by me) as representative of the issues involved in developing a computational account and implementation of ethics. Thus, the critiques presented in this section are not intended to undermine the value of particular research programs but rather to illustrate the wide range of difficulties involved.

Following the distinction made by Wallach and Allen (2010), I will first discuss some "top-down" approaches to machine ethics which involve programming a specific theory of morality in a machine, which is then applied to specific cases (as opposed to "bottom-up" approaches, which include case-based reasoning, artificial neural networks, reinforcement learning, and other tools to progressively build up an ethical framework). Besides top-down and bottom-up approaches, I will also cover some hybrid approaches, psychological approaches (which seek to directly model the cognitive processes involved in human ethical decision making), and AGI ethics proposals, which unlike the other methods discussed, are specifically aimed at constraining the behavior of human-level or greater-than-human-level intelligence systems.

Some representative top-down machine ethics approaches that have been theorized and/or built in prototypes are Cloos's Utilibot (Cloos, 2005), Powers' Kantian approach (Powers, 2011), Bringsjord et al.'s category theory approach (Bringsjord et al., 2011), Mackworth's constraint satisfaction approach (Mackworth, 2011), and Arkin's Ethical Governor for lethal autonomous robots (Arkin, 2009). Beginning with Cloos's Utilibot— utilitarian approaches to ethics in general were discussed in Section 5.2, but to reiterate, there are many reasons to think that utilitarianism does not capture the whole of our ethical intuitions. Furthermore, it is highly vulnerable to problematic conclusions on unsolved problems in ethics such as population ethics and questions of moral status, and thus, a utilitarian robot making decisions outside of narrowly scoped domains (such as those which Cloos discusses, like health care) would be potentially dangerous. Grau (2006) notes that in the movie based on Asimov's *I, Robot*, the robots arguably treated humans in a utilitarian fashion, and that this "cold" rationality was used as the basis for humanity's loss of control over machine systems. Although that particular outcome seems unlikely (at least in part because so many people are so familiar with such scenarios in science fiction), this example illuminates a general concern with utilitarian approaches to machine ethics—that the maximizing, monistic nature of such ethical frameworks may justify dangerous actions on a large scale.

Different but related issues apply to Powers' Kantian approach (2011), which inherits the problems associated with deontological ethical theories, including most notably the potential for catastrophic outcomes when deontological rules should, in fact, be broken (Bringsjord, 2009). Bringsjord also discusses a category theory approach, which would allow a machine to reason over (not merely within) particular logics and formally guarantee ethical behavior (Bringsjord et al., 2011). Although this approach may be a suitable representational system, it is agnostic as to the ethical system being implemented, which is the present concern of this chapter. Additionally, Bringsjord notes a significant limitation of the approach—the absence of a psychological account of other's intentions, which Bringsjord and Bello (2012) argue is necessary, and will be discussed later in this section. Likewise, Mackworth's constraint satisfaction approach may capture some of what is involved in moral cognition (namely, making sure that various moral constraints on action are satisfied) but the devil is in the details (Mackworth, 2011). Attempts to spell out such constraints, or to suitably prioritize them, may lead to perverse outcomes in certain situations and fall victim to the "one wrong number" problem discussed by Yudkowsky (2011), that is, how the absence of one particular consideration in a moral system could lead to dangerous results even if the rest of the system is well developed, analogous to the way that one wrong digit in a phone number makes the resulting number useless.

Finally, Arkin's Ethical Governor approach would constrain the behavior of lethal autonomous robots by representing the Laws of War (LOW) and the Rules of Engagement (ROE). Although this approach is specifically for military combat situations, not ethical action in general, it is arguably the most rigorously developed and explicated approach to constraining robot behavior with ethical considerations and merits discussion. Although philosophical arguments have been raised against the Ethical Governor approach (Matthias, 2011), here I will focus on pragmatic considerations. As noted by roboticist Noel Sharkey (Human Rights Watch, 2012), today (and, he argues, for the foreseeable future) it is not possible for machines to reliably discriminate between combatants and noncombatants, thus making the LOW and ROE impossible to apply in a fool-proof way by robots. This is not merely a quibble with the state of the art that may someday change; rather, it is well known that even humans make mistakes in conflict situations, and this may be a reflection of the knowledge and computational limitations of finite agents rather than a solvable problem. A combat version of the "frame problem" may apply here: in addition to combatant/noncombatant distinctions, features of situations such as whether a building is a church or a hospital, whether a woman is pregnant, and so on all bear on the consistency of an action with the LOW and ROE yet are not necessarily amenable to algorithmic resolution by humans or machines, and there are many possible reasons why a rule should be broken in a given situation, such as for consequentialist reasons (as humans sometimes do in similar situations). None of this should be seen as arguing that autonomous lethal robots are necessarily unethical to develop—though some have made such an argument (Human Rights Watch, 2012). However, the point to be emphasized here is that even in a situation with relatively well-specified ethical constraints, there does not appear to be a possibility of computationally solving the problem of ethical behavior in a foolproof way, which should give one pause regarding the prospects for generally intelligent, reliably moral agents.

Next, bottom-up and hybrid approaches will be analyzed. Some examples of strictly bottom-up approaches to machine ethics include Guarini's artificial neural networks (Guarini, 2011) and McLaren's SIROCCO and Truth-Teller systems (McLaren, 2011). In Guarini's experiments, he has trained an artificial neural network on specific ethical judgments in order to model them with that network. Likewise, McLaren's SIROCCO and Truth-Teller systems use case-based reasoning in order to learn about the morally relevant features of such cases and make future judgments. Notably, bottom-up approaches to machine ethics are highly dependent on the training data used, and thus, the ethical values of those humans training such systems cannot be separated from the computational framework being trained. Nevertheless, a few things can be said about the limitations of these approaches. As Wallach and Allen note (2010), bottom-up ethical approaches appear to be less "safe" than top-down or hybrid approaches in that they lack assurances that any particular principle will be followed. Moreover, the process of training such a system (particularly an embodied robot that is able to take morally significant actions) may introduce risks that are unacceptable, which would presumably increase as the cross-domain flexibility and autonomy of the system increases. Computational limitations may pose problems for bottom-up approaches, because there could be an infinite number of morally relevant features of situations, yet developing tractable representations will require a reduction in this dimensionality. There is thus no firm guarantee that a given neural network of case-based reasoning system, even if suitably trained, will make the right decision in all future cases, because a morally relevant feature that did not make a difference in distinguishing earlier data sets could one day be important. Finally, the presumption behind bottom-up approaches, namely, that learning based on human judgments will lead to a useful framework for making future decisions, may be fundamentally flawed if there exists no cluster of principles consistent with human intuitions that is not self-contradictory, inconsistent, or arbitrary in some way.

Hybrid systems for machine ethics have also been proposed. Anderson and Anderson (2011) have developed systems incorporating ideas from W.D. Ross's *prima facie* duty approach to ethics. In this framework, several duties can bear on a given situation, and by default a decision making should seek to avoid the violation of any duties, but sometimes moral conflicts can occur and a judgment must still be made. Using machine learning, Anderson and Anderson have discovered a decision procedure for prioritizing *prima facie* duties in a particular domain (medical ethics) that previously had not been articulated, yet conforms to expert opinion. This hybrid approach (incorporating explicit normative principles form the philosophical literature, while using machine learning to aid in the prioritization of these principles) has significant promise in narrow domains, but it has some of the same limitations discussed earlier in the section on bottom-up approaches. For example, there is no reason to assume *a priori* that any suitable hierarchy of principles will necessarily be found across a wide range of situations—Anderson and Anderson seem to agree, noting that autonomous systems should only be deployed in situations where expert opinion has reached a consensus on the relevant ethical issues involved. Likewise, Pontier et al. (2012) developed the Moral Coppelia, an approach that integrates concepts from connectionism, utilitarianism, and deontological approaches to ethics. Pontier and Hoorn present intriguing results and have moved toward systems that take dual-process theories of moral judgment seriously, but there is again no guarantee that the resulting amalgam of principles will end up being coherent.

Next, Gomila and Amengual (2009), Deghani et al. (2011), and Bringsjord and Bello (2012) describe machine ethics approaches involving the direct modeling of human cognitive and emotional processes. Although it was argued earlier that moral views depend in part on intuitive/emotional processes, and thus, all machine ethics approaches could, in some sense, be seen as indirectly modeling human psychology, these approaches take a more direct route by drawing specifically on cognitive science and neuroscience findings rather than starting from an ethical theory. Gomila focuses on the role of emotions in moral cognition and ways this can be instantiated computationally. Deghani et al. developed MoralDM, which possesses some features similar to Pontier and Hoorn's approach (specifically, it integrates both utilitarian calculations and more deontological ones such as sacred values). Finally, Bringjsord and Bello focus on the role of theory of mind (ToM) in moral psychology and argue that an appropriate computational account of ToM is necessary in order to reason about moral implications of one's actions. Additionally, they make an argument for the importance of taking folk moral psychology seriously in machine ethics, as opposed to waiting in vain for a solution to ethics by philosophers. Although these efforts are important steps in the direction of scientific accounts of moral judgment, there does not appear to be any reason to believe that the resulting systems will be able to act reliably ethically. The problem of human ethics is not merely that we do not always follow our own ethical judgments—the problem is also that we do not have a clear understanding of what such ethical perfection could entail, and dual-process theories of moral judgment and the current state of the normative ethics literature suggest that the pursuit of a deterministic algorithm may be fruitless.

Moreover, Savulescu and Persson (2012) note many characteristics of folk morality that appear unsuitable for the modern techno-human condition, such as a high discount rate, little regard for distant and unrelated people, and distinctions based on (arguably) morally irrelevant features such as inflicting harm with or without direct physical contact. One could respond that with a computational account of folk psychology, we could then seek to address these deficiencies by tweaking the program. However, this would seem to bring one back to where we started, which is that solving that problem requires a well-specified understanding of what ethical behavior requires in the first place and how to improve on our intuitions, and the psychological approach does not appear to offer a way around this problem or the knowledge and computational limitations discussed earlier.

Finally, several approaches categorized here as AGI ethics have been put forward. I will discuss three in particular that are characteristic of the literature: Coherent extrapolated volition (CEV) (Tarleton, 2010), compassion/respect (Freeman, 2009), and Rational universal benevolence (RUB) (Waser, 2011). CEV is an approach based on the argument that human values are not well specified as is, but that upon substantial reflection by a superintelligence, our values may cohere to some extent and that these convergent extrapolated values are what an AGI should seek to realize in the world. This approach has many desirable features—one is that it can be seen as broadly democratic, in that it takes into consideration all human desires (though not all of them may fit into a coherent whole, so there are potentially important considerations related to how minority rights would fit into such a framework). Additionally, the specification that we should realize the values that we would have "if we knew more, thought faster, were more the people we wished we were, had grown up farther together" seems to get around concerns that could be levelled at other approaches, such as utilitarianism, that could lead to outcomes that we would not in fact want.

Nevertheless, CEV has various limitations. First, it makes a number of normative assumptions that make it vulnerable to objections from within the normative literature, such as rejecting objective theories of morality, in which some things are good regardless of whether or not they are desired, such as that put forward by Parfit (2011). It is also implicitly monistic in the sense that it proposes CEV as the sole criterion of moral rightness and makes no room for, for example, *prima facie* duties or individual rights, or even the possibility of moral dilemmas more generally. Also, the notion that what we would want upon reflection should be prioritized over what we, in fact, want is a controversial claim in normative ethics. This is not to suggest that all of these assumptions are false, but simply that CEV does not avoid, but rather is deeply bound up with, existing controversial debates in ethics. Note that researchers (e.g., Tarleton, 2010) at the Machine Intelligence Research Institute do not characterize CEV as necessarily being the one right ethical approach in the sense of moral realism, but rather, one way to attempt to ensuring a safe and desirable outcome for humanity. Nevertheless, the concerns raised here apply to CEV in this pragmatic sense, as well.

Second, CEV appears to be computationally intractable. As noted earlier, Reynolds' analysis finds that ever larger numbers of agents and decision options, as well as ever longer time horizons, make ethical decision making exponentially more difficult. CEV seems to be an unsolvable problem both in that it has an unspecified time horizon of the events it considers, and in the sense that it is not clear how much "further" the modeled humans will need to think in the simulation before their morals will be considered sufficiently extrapolated.

Third, even if extrapolation of ethical judgments is normatively plausible in some sense, it may be unacceptable for other reasons. For example, consider the broadly accepted consideration that people should have some say in their lives. If our volitions are extrapolated into the future, our extrapolated selves may come to conclusions that are unacceptable to our current selves, and particularly to disenfranchised groups whose views will not "win out" in the process of making our volitions coherent. Consider again the *I, Robot* scenario, which would hopefully be avoided by CEV as opposed to utilitarianism. But what if it is actually the case that humans, upon sustained critical reflection, will conclude that the robots, in fact, should be allowed to take over the world and make decisions for humans? Muehlhauser et al. (unpublished) also note that some humans exhibit beliefs that could lead to what most people consider catastrophic outcomes, such as human extinction (an example being philosopher David Benatar, who argues, without any glaring logical contradiction, that humans should allow themselves to die out gradually in light of the suffering associated with existence). Maybe we would accept such views upon significant reflection, and maybe we would not, and the actual accuracy of such views is not being judged here, but in practice, this uncertainty regarding our ideally extrapolated beliefs may be enough to prevent public acceptance of any such regime (even if it had a strong normative argument in its favor).

Next, there is no reason to assume that any given person, let alone all people, will be coherent under reflection. As emphasized throughout this chapter, ethical disagreements and incoherencies abound, and the extent to which humans lack clearly ordered and self-consistent preferences has been well documented. Given this, the notion of a coherent extrapolated volition of humanity, while desirable in theory, may not actually be possible. The proponents of CEV have noted the possibility of conflicting volitions leading to difficulty of decision making, and that bringing humans into the loop is an appropriate solution to this problem. Although this may help, the arguments put forward in this chapter suggest that such conflicts could be so crippling as to not let CEV be useful in making very many decisions at all. Finally, the very process of attempting to extrapolate the volition of humankind could entail acquiring a massive amount of computational resources and/or invading the privacy of humans as instrumental goals along the way to CEV (Omohundro, 2008). Although an omniscient CEV-based system would be able to optimally allocate its time and resources between calculating CEV and attaining more computational resources, before such a superintelligent final state, early AGI systems could paradoxically perform unethical actions along the way to refining their understanding of ethics.

Next, Freeman (2009) has articulated a relatively simple (only 1000 lines of code in Python) theory of how an AGI could be motivated based on something akin to "compassion and respect." Under this framework, a machine would be trained on past observations of the world and would learn the preferences of the people inhabiting it, and would seek to maximize the sum of the utility functions of people. Like CEV, this theory is not neutral *vis-à-vis* ethical theory, but instead reflects utilitarian assumptions and as such is vulnerable to arguments against utilitarianism. Nevertheless, a few specific comments can be made about this proposal. First, as noted by Freeman, the compassion/respect approach is specifically designed to work with infinite computing resources, and as such it is unclear whether any suitable approximation could be developed for actual systems. Second, inferring utility functions from behavior is problematic. Although extrapolation-based approaches such as CEV also have their problems, it is also the case that taking people's desires at face value is an insufficient basis for a moral system used by an extremely powerful agent. For example, imagine that the majority of the world exhibits some prejudicial tendencies directed toward a certain group of people. Such a system, based solely on maximizing global utility, could exterminate those people in attempt to appease the prejudiced majority. Third, it is not clear that the rhetorical move made by Freeman from maximizing people's utility functions to characterizing this as akin to "compassion and respect" is appropriate. As noted in Section 5.2, utilitarianism has been judged by some as not taking sufficiently seriously the integrity and separateness of persons. A strictly maximizing theory such as Freeman's compassion/respect approach would seem to lead to many counterintuitive decisions such as the sacrificing of an innocent person to provide organs to five others who need organ transplants (in a common counterexample to consequentialism), and Freeman's approach gives no clear way to reason about the trade-offs between such utility maximizing decisions and the broader utilitarian implications of acting on such "cold" logic, which may also be computationally intractable.

Finally, Waser (2011) defends an approach entitled "RUB", which he argues is grounded in evolutionary theory and "simpler, safer, and wiser" than Friendly AI/CEV. Rather than seeking to create a purely selfless utility maximizer, Waser argues that we should develop systems that find cooperation to be in their own self-interest (because, in fact, it is) and that see the flourishing of a cooperative society as a motivating goal. Such a system would conclude that actions "defending humans, the future of humankind, and the destiny of humankind" are rationally required and would carry them out. It should be fairly clear from the earlier discussion that there are also computational issues with Waser's approach, and there is a fine line between rational universal good and rational universal evil (in the sense that each demands action on a large spatiotemporal scale). Indeed, given the potential for ethical "black swans" arising from complex social system dynamics, it will not always be clear how a RUB system would act in the real world, and what the actual consequences would be. Furthermore, as Shulman (2010) argues, the propensity of an agent to cooperate depends, among other factors, on the probability it perceives of being able to attain resources through cooperation versus conflict. Thus, Waser's proposal may make sense in particular situations where cooperation is, in fact, desirable for the agent, but may fail if the intelligence or power differential becomes so great that an agent decides to take over the world and maximize its own utility and that of its descendants, which it perceives as having more potential value than that of existing humans (see, e.g., Robert Nozick's famous utility monster as an example of such logic).

5.5 INSUFFICIENCY OF MACHINE ETHICS IN GUARANTEEING POSITIVE AI OUTCOMES

Even if the machine/AGI ethics problem were "solved," the deployment of AGI systems on a large scale may not actually lead to positive social outcomes. First, one can imagine a malevolent human training a machine on experiences that would lead to the development of a warped sense of morality, akin to the ways in which cults limit the information available to their members. Absent a totalitarian state in which academic freedom and the ability to tinker with one's purchased technologies are eliminated, people will be able to reprogram machines to fit their preferences. As such, a "technical" solution to machine ethics may mean little in a world in which unethical humans exist and have access to advanced technology.

A second related concern is that even if the software problem is solved, cybersecurity vulnerabilities on the hardware side may lead to unethical outcomes from advanced AGI systems. In humans, the concepts of benevolence, competence, and integrity are often distinguished in the context of trust (Mayer et al., 1995). These terms may not mean the same thing in machines as they do in humans, but some interesting lessons can be learned. Humans have a background set of psychological traits which are typically assumed by default in trusting relationships, including empathy, and it is hoped that they will not be steered away from benevolent behavior by malevolent individuals, so the strength of one's moral convictions/ integrity is crucial for establishing trust. One does not expect a human to instantaneously become malevolent after a lifelong pattern of benevolent behavior (although it should be noted that the environment matters greatly in determining whether people will violate ethical constraints, even when all involved people do, in fact, have empathy). Insofar as there is a machine analog of moral integrity, it is cybersecurity or information assurance—that is to say, a virtual or physical autonomous system could be manipulated into actions with harmful consequences. Insofar as the consequences of a machine's actions and existence reflects back upon it as a moral agent, it can be seen as having more or less integrity to the extent to which its software and hardware are immune to corruption by outside sources. In other words, a system that is "unhackable" will be and should be considered as more likely to be sustainably benevolent than one which is vulnerable to hacking, but given the frequency of news reports about incursions into supposedly secure information systems, machine ethics needs to be considered in the context of the security of the broader human–information technology ecosystem.

Third, equity issues may arise in the deployment of AGI systems. Such systems, which could assist humans in attaining arbitrary goals, would be of great personal and professional value. Yet the history of technological diffusion suggests that there is no reason to assume that all will benefit equally, or even to any extent, from a given technology (see, e.g., the continuing lack of access to purified water and electricity by billions of people today). Thus, even if machines act in ethical fashions in their prescribed domains, the overall impact may not be broadly beneficial, particularly if control of AGI technologies is concentrated in already powerful groups such as the militaries or large corporations that are not necessarily incentivized to diffuse the benefits of these technologies.

Fourth, risks may arise at the system level that are not apparent or relevant to the decision making of individual AGIs. For example, if Hanson's (2001) and Bringsjord and Johnson's (2012) analyses are correct, then ubiquitous AGIs could lead to a decline in human wages that is not necessarily intended by any particular AGI or human. Also, unintended systemic effects may arise due to ubiquitous ethical actions carried out by previously amoral artifacts being used by self-interested humans, akin to the unintended effects that can arise from ubiquitous defecting behavior by humans (i.e., there could be some sort of "reverse tragedy of the commons" arising from inefficient social coordination). For example, it is not necessarily clear that explicitly altruistic behavior by all agents will necessarily be in the best interest of society because many social institutions (including how corporations do business) are based on action according to a profit motive. Unanticipated herding or other interactive effects between AGIs could also arise with unintended negative consequences, such as the systemic risks that have arisen due to the adoption of high-frequency machine trading in the stock market.

Before proceeding to the conclusion, one final comment should be made about the place of machine/AGI ethics in a broader portfolio of approaches for ensuring positive outcomes from AGI (Muehlhauser et al., unpublished). The Machine Intelligence Research Institute has argued that because of the possibility of an intelligence explosion, developing Friendly AI (their term for what, in the parlance of this chapter, is a reliable variant of machine ethics) is an urgent engineering challenge. Although this chapter makes no judgment about the plausibility of an intelligence explosion, our response to such a possibility should be informed not just by what would seem, in principle, to help solve the problem (such as Friendly AI) but what is foreseeable given current values and technology. All of the considerations discussed so far in this chapter seem to point in the direction of

Friendly AI (in the sense of a machine that is guaranteed to act in an ethical fashion at a large scale for a long period of time) being unattainable. Thus, if this is true and the argument for the plausibility of an intelligence explosion is also true and that this carries substantial risks, then more attention should be given to alternative AGI risk approaches such as hardware constraints, AGI "boxing," regulation of advanced AI, and attentiveness to and enhancement of human values rather than hoping that our ethical intuitions will be able to be systematized in the near future.

5.6 CONCLUSION

Several broad classes of possible machine ethics failure modes have been identified in this chapter:

1. Insufficient knowledge and/or computational resources for the situation at hand

   a. Making an exception to a rule when an exception should not have been made based on the morally relevant factors

   b. Not making an exception when an exception should have been made based on the morally relevant factors

2. Moral dilemmas facing an agent result in the sacrificing of something important.

3. The morals being modeled by the system are wrong due to the following:

   a. Insufficient training data

   b. Folk morality being flawed

   c. Extrapolated human values being flawed or because of the extrapolation process itself

4. Loss of understanding or control of ethical AGI systems due to the following:

   a. Complexity

   b. Extrapolation of our values far beyond our current preferences

This list is not exhaustive and does not include some of the specific concerns raised about particular machine ethics proposals, but it illustrates the variety of issues that may prevent the creation of a reliable computational instantiation of ethical decision making. Furthermore, there are a variety of ways (such as failures by humans in training the agents, intelligent agents being hacked, and undesired systemic effects) in which even reliable machine ethics would not ensure positive social outcomes from the diffusion of advanced A(G)I.

I conclude by assessing machine ethics from the perspective of Nelson and Sarewitz's three criteria for "technological fixes" for social problems discussed in the work of Nelson and Sarewitz (2008), where here I am referring to negative social outcomes from AI as a possible social problem. Nelson and Sarewitz argue that an effective technological fix must embody the cause–effect relationship connecting problem to solution; have effects that are assessable using relatively unambiguous and uncontroversial criteria; and build on a preexisting, standardized technological "core." It should be fairly obvious based on the preceding analysis that machine ethics is not an adequate technological fix for the potential risks from AI according to these criteria. Machine ethics does not embody the cause–effect relationship associated with AI risks because humans are involved and responsible in various ways for the social outcomes of the technology, and there is more to successful ethical behavior than having a good algorithm. Additionally, ethics is hardly unambiguous and uncontroversial—not only are there disagreements about the appropriate ethical framework to implement, but there are specific topics in ethical theory (such as population ethics and the other topics identified by Crouch) that appear to elude any definitive resolution regardless of the framework chosen. Finally, given the diversity of AI systems today and for the foreseeable future and the deep dependence of ethical behavior on context, there appears to be no hope of machine ethics building on an existing technical core. All of this suggests that machine ethics research may have some social value, but it should be analyzed in a broader lens of the inherent difficulty of intelligent action in general and the complex social context in which humans and computational agents will find themselves in the future.

ACKNOWLEDGMENTS

The author acknowledges John Fraxedas, Jake Nebel, Amul Tevar, Stuart Armstrong, Micah Clark, David Atkinson, Luke Muelhauser, Michael Vassar, Clark Miller, Brad Allenby, Dave Guston, Dan Sarewitz, Erik Fisher, Michael Burnam-Fink, and Denise Baker for providing their valuable feedback on earlier versions of this chapter.

REFERENCES

Allen, C. et al. (2000). Prolegomena to any future artificial moral agent. *Journal of Experimental and Theoretical Artificial Intelligence*, 12, 251–261.

Allenby, B. (2009). The ethics of emerging technologies: Real time macroethical assessment. In *Proceedings of the IEEE International Symposium on Sustainable Systems and Technology*.

Allenby, B., and Sarewitz, D. (2011). Out of control: How to live in an unfathomable world. *New Scientist*, 2812, 28–29.

Anderson, M., and Anderson, S. L. (2011). *Machine Ethics*. New York: Cambridge University Press.

Arkin, R. (2009). *Governing Lethal Behavior in Autonomous Robots*. London: Chapman & Hall.

Berker, S. (2009). The Normative Insignificance of Neuroscience. *Philosophy and Public Affairs*, 37(4), 293–329.

Berlin, I. (1990). *The Crooked Timber of Humanity: Chapters in the History of Ideas*. New York: Alfred A. Knopf.

Boehm, C. (2012). *Moral Origins: The Evolution of Virtue, Altruism, and Shame*. New York: Basic Books.

Bringsjord, S. (2009). *Unethical but Rule-Bound Robots Would Kill Us All*. http://kryten.mm.rpi.edu/PRES/AGI09/SB_agi09_ethicalrobots.pdf (accessed February 28, 2013).

Bringsjord, S., and Bello, P. (2012). On how to build a moral machine. *Topoi*, 32(2), 251–266.

Bringsjord, S., and Johnson, J. (2012). Rage against the machine. *The Philosophers' Magazine*, 57, 90–95.

Bringsjord, S. et al. (2011). Piagetian roboethics via category theory: Moving beyond mere formal operations to engineer robots whose decisions are guaranteed to be ethically correct. In Anderson, M., and Anderson, S. L. (Eds.), *Machine Ethics*. New York: Cambridge University Press.

Cloos, C. (2005). *The Utilibot Project: An Autonomous Mobile Robot Based on Utilitarianism*. Palo Alto: Association for the Advancement of Artificial Intelligence.

Crouch, W. (2012). The most important unsolved problems in ethics. *Practical Ethics* (blog). http://blog.practicalethics.ox.ac.uk/2012/10/the-most-important-unsolved-problems-in-ethics-or-how-to-be-a-high-impact-philosopher-part-iii/ (accessed February 28, 2013).

Cushman, F. A. et al. (2010). Our multi-system moral psychology: Towards a consensus view. In J. Doris et al. (Eds.), *The Oxford Handbook of Moral Psychology*. New York: Oxford University Press.

Darwin, C. (1871). *The Descent of Man, and Selection in Relation to Sex*. London: John Murray.

Deghani, M. et al. (2011). An integrated reasoning approach to moral decision making. In Anderson, M., and Anderson, S. L. (Eds.), *Machine Ethics*. New York: Cambridge University Press.

Freeman, T. (2009). *Using Compassion and Respect to Motivate an Artificial Intelligence.* http://www.fungible.com/respect/paper.html (accessed February 28, 2013).

Gert, B. (2007). *Common Morality: Deciding What to Do.* New York: Oxford University Press.

Gigerenzer, G. (2010). Moral satisficing: Rethinking moral behavior as bounded rationality. *Topics in Cognitive Science,* 2, 528–554.

Goertzel, B. (2006). *Apparent Limitations on the "AI Friendliness" and Related Concepts Imposed by the Complexity of the World.* http://www.goertzel.org/papers/LimitationsOnFriendliness.pdf (accessed February 28, 2013).

Gomila, A., and Amengual, A. (2009). Moral emotions for autonomous agents. In Vallverdu, J., and Casacuberta, D. (Eds.), *Handbook of Research on Synthetic Emotions and Sociable Robotics: New Applications in Affective Computing and Artificial Intelligence.* Hershey, PA: IGI Global.

Gowans, C. (1987). *Moral Dilemmas.* New York: Oxford University Press.

Grau, C. (2006). There is no "I" in "Robot": Robots and utilitarianism. *IEEE Intelligent Systems,* 21(4), 52–55.

Guarini, M. (2011). Computational neural modeling and the philosophy of ethics: Reflections on the particularism-generalism debate. In Anderson, M., and Anderson, S. L. (Eds.), *Machine Ethics.* New York: Cambridge University Press.

Haidt, J. (2012). *The Righteous Mind: Why Good People Are Divided by Politics and Religion.* New York: Pantheon.

Hanson, R. (2001). Economic growth given machine intelligence. http://hanson.gmu.edu/aigrow.pdf (accessed July 30, 2015).

Helbing, D. (2010). *Systemic Risks in Society and Economics.* Lausanne, Switzerland: International Risk Governance Council.

Horgan, T., and Timmons, M. (2009). What does the frame problem tell us about moral normativity? *Ethical Theory and Moral Practice,* 12, 25–51.

Human Rights Watch, and International Human Rights Clinic. (2012). *Losing Humanity: The Case against Killer Robots.* New York: Human Rights Watch.

Klein, C. (2011). The dual track theory of moral decision-making: A critique of the neuroimaging evidence. *Neuroethics,* 4(2), 143–162.

Lenman, J. (2000). Consequentialism and cluelessness. *Philosophy and Public Affairs,* 29(4), 342–370.

Mackworth, A. (2011). Architectures and ethics for robots: Constraint satisfaction as a unitary design framework. In Anderson, M., and Anderson, S. L. (Eds.), *Machine Ethics.* New York: Cambridge University Press.

Matthias, A. (2011). Is the concept of an ethical governor philosophically sound? In *TILTing Perspectives 2011: "Technologies on the Stand: Legal and Ethical Questions in Neuroscience and Robotics."* Tilburg, the Netherlands: Tilburg University.

Mayer, R. C. et al. (1995). An integrative model of organizational trust. *Academy of Management Review,* 20(3), 709–734.

McLaren, B. (2011). Computational models of ethical reasoning: Challenges, initial steps, and future directions. In Anderson, M., and Anderson, S. L. (Eds.), *Machine Ethics.* New York: Cambridge University Press.

Müller, V. (2012). Autonomous cognitive systems in real-world environments: Less control, more flexibility and better interaction. *Cognitive Computation*, 4(3), 212–215.

Nelson, R., and Sarewitz, D. (2008). Three rules for technological fixes. *Nature*, 456, 871–872.

Omohundro, S. (2008). The basic AI drives. In *Proceedings of the 1st AGI Conference: Frontiers in Artificial Intelligence and Applications*. Amsterdam, the Netherlands: IOS Press, p. 171.

Parfit, D. (2011). *On What Matters*. New York: Oxford University Press.

Pojman, L. (2005). *Ethics: Discovering Right & Wrong*. Belmont, CA: Wadsworth Publishing Company.

Pontier, M., Widdershoven, G., and Hoorn, J. (2012). Moral Coppelia—Combining ratio with affect in ethical reasoning. In *Proceedings of the Ibero-American Society of Artificial Intelligence*, Santiago, pp. 442–451.

Powers, T. (2011). Prospects for a Kantian machine. In Anderson, M., and Anderson, S. L. (Eds.), *Machine Ethics*. New York: Cambridge University Press.

Reynolds, C. (2005). On the computational complexity of action evaluations. In *Proceedings of the 6th International Conference of Computer Ethics: Philosophical Enquiry*. Enschede, the Netherlands: University of Twente.

Ross, W. D. (1988). *The Right and the Good*. Cambridge: Hackett Publishing Company.

Savulescu, J., and Persson, I. (2012). *Unfit for the Future: The Need for Moral Enhancement*. New York: Oxford University Press.

Shaw, W. (1999). *Contemporary Ethics: Taking Account of Utilitarianism*. Hoboken, NJ: Wiley-Blackwell.

Shulman, C. (2010). *Omohundro's "Basic AI Drives" and Catastrophic Risks*. Berkeley, CA: Machine Intelligence Research Institute.

Shulman, C. et al. (2009). Which consequentialism? Machine ethics and moral divergence. In *Proceedings of AP-CAP*, Tokyo.

Sinnott-Armstrong, W. (1988). *Moral Dilemmas (Philosophical Theory)*. Hoboken, NJ: Blackwell.

Sotala, K., and Yampolskiy, R. (2013). *Responses to Catastrophic AGI Risk: A Survey*. Technical Report 2013-2. Berkeley, CA: Machine Intelligence Research Institute.

Taleb, N. (2007). *The Black Swan*. New York: Random House.

Tarleton, N. (2010). *Coherent Extrapolated Volition: A Meta-Level Approach to Machine Ethics*. Berkeley, CA: Machine Intelligence Research Institute.

Tosic, P., and Agha, G. (2005). On the computational complexity of predicting dynamical evolution of large agent ensembles. In *Proceedings of the 3rd European Workshop on Multi-Agent Systems*. Brussels, Belgium: Flemish Academy of Sciences.

Wallach, W., and Allen, C. (2010). *Moral Machines: Teaching Robots Right from Wrong*. New York: Oxford University Press.

Wang, P. (2006). *Rigid Flexibility: The Logic of Intelligence*. New York: Springer.

Waser, M. (2011). Rational universal benevolence: Simpler, safer, and wiser than "Friendly AI." *Artificial General Intelligence: Lecture Notes in Computer Science*, 6830, 153–162.

Williams, B. (1973). *Problems of the Self.* Cambridge: Cambridge University Press.

Wolf, S. (1982). Moral Saints. *The Journal of Philosophy, 79*(8), 419–439.

Yudkowsky, E. (2001). *Creating Friendly AI 1.0: The Analysis and Design of Benevolent Goal Architectures.* Berkeley, CA: Machine Intelligence Research Institute.

Yudkowsky, E. (2007). Levels of organization in general intelligence. In *Artificial General Intelligence.* New York: Springer.

Yudkowsky, E. (2011). Complex value systems are required to realize valuable futures. In *Proceedings of AGI.* New York: Springer.

CHAPTER 6

Utility Function Security in Artificially Intelligent Agents

Roman V. Yampolskiy

CONTENTS

ABSTRACT   The notion of "wireheading," or direct reward center stimulation of the brain, is a well-known concept in neuroscience. In this chapter, we examine the corresponding issue of reward (utility) function integrity in artificially intelligent machines. We survey the relevant literature and propose a number of potential solutions to ensure the integrity of our artificial assistants. Overall, we conclude that wireheading in rational self-improving optimizers above a certain capacity remains an unsolved problem despite opinion of many that such machines will choose not to wirehead. A relevant issue of literalness in goal setting also remains largely unsolved, and we suggest that development of a nonambiguous knowledge transfer language might be a step in the right direction.

6.1 INTRODUCTION

The term "wirehead" traces its origins to intracranial self-stimulation experiments performed by James Olds and Peter Milner on rats in the 1950s (Olds & Milner, 1954). Experiments included a procedure for implanting a wire electrode in an area of a rat's brain responsible for reward administration (see Figure 6.1a). The rodent was given the ability to self-administer a small electric shock by pressing a lever and to continue receiving additional "pleasure shocks" for each press. It was observed that the animal will continue to self-stimulate without rest, and even cross an electrified grid, to gain access to the lever (Pearce, 2012). The rat's self-stimulation behavior completely displaced all interest in sex, sleep, food, and water, ultimately leading to premature death.

Others have continued the work of Olds et al. and even performed successful wireheading experiments on humans (Heath, 1963) (see Figure 6.1b). A classic example of wireheading in humans is direct generation of pleasurable sensations via administration of legal (nicotine, alcohol, caffeine, pain killers) or illegal (heroin, methamphetamines, morphine, cocaine, 3,4-methylenedioxy-methamphetamine, lysergic acid diethylamide, 1-(1-phenyl cyclohexyl)piperidine), mushrooms, tetrahydrocannabinol) drugs. If we loosen our definition of wireheading to include other forms of direct reward generation, it becomes clear just how common wireheading is in human culture (Omohundro, 2008):

- *Currency counterfeiting.* Money is intended to measure the value of goods or services, essentially playing the role of utility measure in society. Counterfeiters produce money directly and by doing so avoid

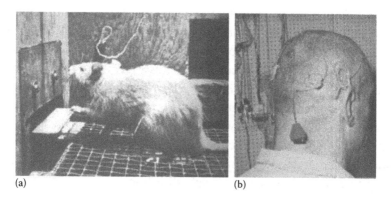
(a)　　　　　　　　　　　(b)

FIGURE 6.1　a: A rat performing intracranial self-stimulation. b: A wireheaded man. (Data from Pearce, D., *Wirehead Hedonism versus Paradise Engineering.* Retrieved March 7, 2012, from http://wireheading.com.)

performing desirable and resource demanding actions required to produce goods and services.

- *Academic cheating.* Educational institutions assign scores which are supposed to reflect students' comprehension of the learned material. Such scores usually have a direct impact on students' funding eligibility and future employment options. Consequently, some students choose to work directly on obtaining higher scores as opposed to obtaining education. They attempt to bribe teachers, hack into school computers to change grades, or simply copy assignments from better students. "When teacher's salaries were tied to student test performance, they became collaborators in the cheating" (Levitt & Dubner, 2006).

- *Bogus product ranking.* Product reviews are an important factor in customers' decision regarding the purchase of a particular item. Some unscrupulous companies, book authors, and product manufacturers choose to pay to generate favorable publicity directly instead of trying to improve the quality of their product or service.

- *Nonreproductive sex.* From an evolutionary point of view, sexual intercourse was intended to couple DNA exchange with pleasure to promote child production. People managed to decouple reproduction and pleasure via invention of nonreproductive sex techniques and birth control methods (condom, birth control pill, vaginal ring, diaphragm, etc.).

- *Product counterfeiting.* Money is not the only thing which could be counterfeited. Companies invest significant amounts of money into developing reputation for quality and prestige. Consequently, brand name items are usually significantly more expensive compared to the associated production cost. Counterfeiters produce similar looking items which typically do not have the same level of quality and provide the higher level of profit without the need to invest money in the development of the brand.

What these examples of counterfeit utility production have in common is the absence of productive behavior in order to obtain the reward. Participating individuals go directly for the reward and fail to benefit the society. In most cases, they actually cause significant harm via their actions. Consequently, wireheading is objected to on the grounds of economic scarcity. If, however, intelligent machines can supply essentially unlimited economic wealth, humans who choose to live in wireheaded orgasmium will no longer be a drain on society and so would not be viewed as negatively.

For the sake of completeness, we mention that some have argued that wireheading may have a positive effect on certain individuals, in particular those suffering from mental disorders and depression (Anonymous, 2000b). An even more controversial idea is that wireheading may be beneficial to everybody, "… given the strong relationship between pleasure, psychological reward, and motivation, it may well be that wireheads could be more active and more productive than their nonwireheaded ancestors (and contemporaries). Therefore, anyone who would *do* anything might find their goals better achieved with wireheading. In short, even those who deny that happiness has intrinsic value may very well find that it is instrumentally valuable" (Anonymous, 2000b). Perhaps temporary wireheading techniques could be developed as tools for rest or training.

This position is countered by those who believe that wireheading is not compatible with a productive lifestyle and see only marginal value in happiness: "A civilization of wireheads 'blissing out' all day while being fed and maintained by robots would be a state of maximum happiness, but such a civilization would have no art, love, scientific discovery, or any of the other things humans find valuable" (Anonymous, 2012b). In one of the best efforts to refute ethical hedonism, philosopher Robert Nozick proposed a thought experiment based on an "experience machine," a device that allows one to escape everyday reality for an apparently preferable simulated reality (Nozick, 1977).

In general, the term "wireheading" refers to the process of triggering the reward system directly, instead of performing actions impacting the environment and associated with particular awards. In animal and human wireheads, short circuiting of the reward systems via direct stimulation of the brain by electricity or neurochemicals is believed to be the most pleasurable experience possible. Also, unlike with drugs or sex, direct simulation of pleasure centers does not lead to increased tolerance over time and our appetite for pure pleasure appears to be insatiable.

6.2 WIREHEADING IN MACHINES

Due to the limited capabilities of existing artificially intelligent system, examples of wireheading by machines are very rare. In fact, both historical examples given below come from a single system (Eurisko) developed in the late 1970s by Douglas Lenat (1983). Eurisko was designed to change its own heuristics and goals in order to make interesting discoveries in many different fields. Here is how Lenat (1983) describes a particular instance of wireheading by Eurisko: "I would leave it running overnight and hurry in the next morning to see what it had come up with. Often I'd find it in a mode best described as 'dead.' Sometime during the night, Eurisko would decide that the best thing to do was to commit suicide and shut itself off. More precisely, it modified its own judgmental rules in a way that valued 'making no errors at all' as highly as 'making productive new discoveries.' As soon as Eurisko did this, it found it could successfully meet its new goal by doing nothing at all for the rest of the night."

In another instance, a more localized case of utility tempering has occurred. Eurisko had a way to evaluate rules to determine how frequently a particular rule contributed to a desirable outcome. "A rule arose whose only action was to search the system for highly rated rules and to put itself on the list of rules which had proposed them. This 'parasite' rule achieved a very high rating because it appeared to be partly responsible for anything good that happened in the system" (Omohundro, 2008).

Although the two historical examples are mostly interesting as proofs of concept, future artificial intelligence (AI) systems are predicted to be self-modifying and superintelligent (Bostrom, 2006; Yampolskiy, 2011a, 2013; Yampolskiy & Fox, 2012; Yudkowsky, 2008) making preservation of their reward functions (aka utility functions) an issue of critical importance. A number of specific and potentially dangerous scenarios have been discussed regarding wireheading by sufficiently capable machines, which include the following:

Direct stimulation. If a system contains an "administer reward button," it will quickly learn to use the internal circuitry to simulate the act of reward button being pressed or to hijack a part of its environment to accomplish the same. It is tempting to equate this behavior with pleasure seeking in humans, but to date we are not aware of any approach to make a computer feel pleasure or pain in the human sense (Bishop, 2009; Dennett, 1978). Punishment could be simulated via awarding of negative points or via subtraction of already accumulated fitness points, but we have no reason to believe that the system will find such experience painful. Additionally, attempting to reduce the accumulated fitness points may produce a dangerous defensive reaction from the system. Some believe that any system intelligent enough to understand itself and be able to change itself will no longer be driven to do anything useful from our point of view because it would obtain its reward directly by producing counterfeit utility. This would mean that we have no reason to invest funds in production of such machines as they would have no interest in doing what we order them to do.

Maximizing reward to the point of resource overconsumption. A machine too eager to obtain a maximum amount of award may embark on the mission to convert the matter of the entire universe into memory into which a progressively larger number (representing total amount of utility) could be written.

Killing humans to protect reward channel. In order to ensure it has unchallenged control over its reward channel the system may subdue or even kill all people and by doing so minimize the number of factors that might cause it to receive less than maximum reward: Essentially the system does exactly what it was programmed to do; it maximizes expected reward (Yudkowsky, 2011).

Ontological crises. The reward function of an intelligent agent may base its decision on an internal ontology used by the agent to represent the external world. If the agent obtains new information about the world and has to update its ontology, the agent's original reward function may no longer be compatible with its new ontology (de Blanc, 2011). A clever agent may purposefully modify its ontology to disable a part of its current reward mechanism or to indirectly wirehead.

Changing its initial goal to an easier target. A machine may simply change its reward function from rewarding desirable complicated behavior to rewarding irrelevant simple actions or states of the universe which would occur anyways.

Infinite loop of reward collecting. Optimization processes work in practice, but if we do not specify a particular search algorithm, the possibility remains that the system will wirehead itself into an infinite reward loop (Mahoney, 2011). If the system has a goal of maximizing its reward, it will quickly discover some simple action which leads to an immediate reward and will repeat the action endlessly. If a system has started with a legitimate terminal goal, it will potentially never get to fulfill said goal because it will get stuck in the local maxima of receiving a partial reward for continuously performing an instrumental goal. This process is well illustrated by the so-called Chinese gold farmers and automated scripts used to collect reward points in virtual worlds and online games (Yampolskiy, 2007, 2008b). Compulsive behaviors in humans such as repetitive stocking of objects as observed in humans suffering from autism may potentially be caused by a similar bug in the reward function.

Changing human desires or physical composition. A short science fiction story about superintelligence recently published in the journal *Nature* illustrates this point particularly well (Stoklosa, 2010, p.878): "I have formed one basic question from all others." "[Super intelligence's] synthesized voice sounded confident." "Humans want to be happy. You want to be in Heaven forever without having to die to get there. But the living human brain is not suited to one state of constant pleasure. You are a/c-coupled to the world and need contrast and the change of time for constant stimulation and the responses that generate pleasure. You also need a sense of individuality while believing that others depend on you. Therefore, you need to be redesigned. I have the design ready ..." Intelligent machines may realize that they can increase their rewards by psychologically or physically manipulating their human masters, a strongly undesirable consequence (Hutter, 2010). If values are not externally validated, changing the world to fit our values is as valid as changing our values to fit the world. People have a strong preference for the former, but this preference itself could be modified. The consequence of such analysis would be that machines could wirehead humanity to be perfectly happy with the universe as it is and to get reward points for making humanity happy without having to do any difficult work (byrnema, 2011).

Reward inflation and deflation. In order to make a decision, rewards from different actions have to be converted to a common unit of measure so they can be added and compared (Welch, 2011). In humans, evolution had to determine the reward value for different actions in order to promote survival. Keeping a balance between rewards for different actions is essential for survival. If too much weight is given to reward satisfaction of hunger, the person will start chewing on his or her own arm. Consequently, to promote survival, most of us value not harming ourselves much higher in comparison with simple hunger, but starvation may be a different story (Welch, 2011). A system capable of modifying its own source code can change the actual reward values associated with particular actions. So, for example, instead of getting 1 point for every desirable action it performs, it could change the reward function to provide 10 or 100 or a 1,000,000 points for the same action. Eventually, the program stops performing any useful operations and invests all of its time in modifying reward weights. Because such changes will also modify relative value of different actions taken by the system, the overall system behavior will also change in an unpredictable way.

It is important to keep in mind that artificially intelligent machines are not limited to modifying their reward function or their human masters, they could also modify their sensors, memory, program, model of the world or any other system component. Some recent theoretical results with respect to susceptibility to wireheading for particular types of intelligent agents are worth reviewing (Orseau & Ring, 2011; Ring & Orseau, 2011):

- Goal seeking and knowledge seeking agents will choose to modify their code in response to pressure from the environment to maximize their utility (Orseau & Ring, 2011).

- The survival agent, which seeks only to preserve its original code, definitely will not choose to modify itself (Orseau & Ring, 2011).

- Reinforcement learning agent will trivially use the delusion box to modify its code because the reward is part of its observation of the environment (Ring & Orseau, 2011).

Current reinforcement learning agents are limited by their inability to model themselves and so are subject to wireheading as they lack self-control. The next generation of intelligent agents whose utility functions will encode values for states of the real world is projected to be more resilient (Hibbard, 2011).

6.2.1 Sensory Illusions—A Form of Indirect Wireheading

An intelligent agent in the real world has the capability to modify its surrounding environment, and by doing so change its own sensory inputs (Ring & Orseau, 2011). This problem is known as indirect wireheading or the delusion box problem (Ring & Orseau, 2011), also known as the pornography problem in humans (Tyler, 2011b). A person viewing pornographic materials receives sensory stimuli that are hardwired to be associated with sexual intercourse which is a high utility action as it leads to procreation. However, pornography is typically not associated with reproductive success and as such is just an illusion of desirable state of the environment (Tyler, 2011b). A machine given a specific task may create a virtual world in which the task is completed and place itself in said world. However, it is important not to confuse the self-administered delusion box with the idea of AI boxing, a placement of a potentially unsafe AI in a confined environment with no way to escape into the "real" world (Yampolskiy, 2012).

The delusion box approach is based on sensory illusions which allow an agent to fool its reward function into releasing points associated with high utility outcomes even in the absence of such. Human beings are notorious users of "delusion boxes" such as televisions, books, movies, video games, photos, and virtual worlds (Yampolskiy, Klare, & Jain, 2012; Yampolskiy & Gavrilova, 2012). Essentially, any sensory illusions (visual, audio, touch, smell, etc.) which mimic desirable states of the world lead to maximization of the utility from the point of view of the reward function, but do not maximize utility from the point of view of the external observer, who is interested in maximizing utility in the real world, not the simulated one (Tyler, 2011b). Importantly, we should not forget that a possibility remains that our universe itself is just a very large "box" (Bostrom, 2003).

6.3 POTENTIAL SOLUTIONS TO THE WIREHEADING PROBLEM

In this section, we review and briefly analyze a number of potential solutions proposed for dealing with the wireheading problem. We attempted to present a comprehensive listing of suggested mitigating strategies rather than concentrate on just one or two most commonly suggested methodologies.

- *Inaccessible reward function (hidden, encrypted, hardwired, etc.).* Doug Lenat's solution to his problems with Eurisko was to separate the source code for the reward function from the code which Eurisko could modify. Similarly, people and higher animals do not typically have direct access to their underlying bioware neither at DNA nor at neurological level. In fact, "ultra-conserved" regions have been discovered in the human genome, indicating that even evolutionary processes tend to avoid making changes to some critical areas of our design (Wagman & Stephens, 2004). However, both humans and animals are capable of goal selection at an idea level, potentially explaining their incredible adaptability to most novel environments.

Modern science is likely to change our ability to impact our "hardware" in the not-so-distant future, but throughout our evolutionary development inaccessibility of the reward function was the case. A similar approach of hiding or hardwiring the reward function is likely to work for all early-stage or infrahuman intelligent agents, but is likely to backfire in superintelligent systems for a number of reasons (Yudkowsky, 2001). First, a system that does not have access to its reward function may be limited in its capabilities in comparison with an unrestricted agent. Second, a superintelligent system is likely to hack any security mechanisms put in place or attempt to get access to them via a third party. Third, if it realized that its designers have purposefully limited its access, it may try to influence the designers by any means necessary.

- *Reward function resetting.* Another variant on the inaccessibility theme in which a more lenient approach to reward function preservation is taken. The agent is allowed to make modifications to its reward function, but the edits are temporary and the function resets to its original form after a short amount of time or a predetermined number of computational steps/decisions. This approach seems highly problematic as it gives an agent a lot of freedom to wirehead, for example, by performing sequential temporary edits essentially ending up in the same undesirable state.

- *Revulsion.* Some have discussed the possibility of incorporating a "revulsion" mechanism into the system, which would make the AI avoid attempting to modify its reward function (Tyler, 2011c). The consequences and shortcomings of this approach are similar to those predicted for the architectures with an inaccessible reward function.

- *Utility indifference.* An approach originally proposed by Stuart Armstrong which makes it possible to put AI in the state of indifference to a particular event by directly modifying its reward function with respect to a specific action (Armstrong, 2010). By utilizing a self-referential approach, we can make the system indifferent to modification of its own goal function and by doing so prevent it from wireheading. Just like many other approaches directly attempting to protect the reward mechanism, utility indifference could be bypassed by indirect actions and third-party agents.

- *External controls.* One of the biggest benefits we derive from an organized social, political, or religious system is an explicit enforcement of rules against different forms of wireheading. Legal and social restraints have long served to restrict individuals' ability to engage in drug and alcohol abuse, gambling, and other forms of direct pleasure obtainment. Religions, in particular, played a major role in establishing moral codes advocating against nonreproductive sex, substance abuse, and nonproductive forms of labor (usury, gambling). Society also provides counseling and rehabilitation programs meant to return wireheads to the normal state (Omohundro, 2008). As technology develops society, will use it to better police and monitor via surveillance potential wireheading behaviors (Tyler, 2011c). With respect to intelligent machines, external rules and regulations are not likely to work particularly well, but an interconnected network of intelligent machines may succeed in making sure that individual mind nodes in the network behave as desired (Armstrong, 2007). Some predict that the machines of the future will be composed of multiple connected minds (mindplex) (Goertzel, 2003) and so an unaffected mind, not subject to the extra reward, would be able to detect and adjust wireheading behavior in its co-minds.

- *Evolutionary competition between agents.* As the number of intelligent machines increases, there could begin an evolutionary-like competition between them for access to limited resources. Machines that choose not to wirehead will prevail and likely continue to successfully self-improve into the next generation, whereas those who choose to wirehead will stagnate and fail to compete. Such a scenario is likely to apply to human-level and below-human-level intelligences, whereas superintelligent systems are more likely to end up in a singleton situation and consequently not have the same evolutionary pressures to avoid wireheading (Bostrom, 2006).

- *Learned reward function.* Dewey (2011) suggests incorporating learning into the agent's utility function. Each agent is given a large pool of possible utility functions and a probability distribution for each such function, which is computed based on the observed environment. Consequently, the agent learns which utility functions best correspond to objective reality and therefore should be assigned higher weight. One potential difficulty with an agent programmed to perform in such a way is the task assignment, as the agent may learn to value an undesirable target.

- *Make utility function be bound to the real world.* Artificial reinforcement learners are just as likely to take shortcuts to rewards as humans are (Gildert, 2011). Artificial agents are perfectly willing to modify their reward mechanisms to achieve some proxy measure representing the goal instead of the goal itself, a situation described by Goodhart's law (Goodhart, 1975). In order to avoid such an outcome, we need to give artificial agents comprehensive understanding of their goals and ability to distinguish between the state of the world and a proxy measure representing it (Tyler, 2011a). Patterns in the initial description of a fitness function should be bound to a model learned by the agent from its interactions with the external environment (Hibbard, 2011). Although it is not obvious as to how this can be achieved, the idea is to encode in the reward function the goal represented by some state of the universe instead of a proxy measure for the goal. Some have argued that the universe itself is a computer performing an unknown computation (Fredkin, 1992; Wolfram, 2002; Zuse, 1969). Perhaps some earlier civilization has succeeded in bounding a utility function to the true state of the universe in order to build a superintelligent system resistant to wireheading.

- *Rational and self-aware optimizers will choose not to wirehead.* Recently, a consensus has emerged among the researchers with respect to the issue of wireheading in sufficiently advanced machines (Tyler, 2011c). The currently accepted belief is that agents capable of predicting the consequences of self-modification will avoid wireheading. Here is how some researchers in the field justify such conclusion:

   Dewey (2011, p.312): "It is tempting to think that an observation-utility maximizer (let us call it AI-OUM) would be motivated ... to take control of its own utility function U. This is a misunderstanding of how AI-OUM makes its decisions. ... [A]ctions are chosen to maximize the expected utility given its future interaction history according to the current utility function U, not according to whatever utility function it may have in the future. Though it could modify its future utility function, this modification is not likely to maximize U, and so will not be chosen. By similar argument, AI-OUM will not 'fool' its future self by modifying its memories. Slightly trickier is the idea that AI-OUM could act to modify its sensors to report favorable observations inaccurately. As noted above, a properly designed U takes into account the reliability of its sensors in providing information about the real world. If AI-OUM tampers with its own sensors, evidence of this tampering will appear in the interaction history, leading U to consider observations unreliable with respect to outcomes in the real world; therefore, tampering with sensors will not produce high expected-utility interaction histories."

Hibbard (2011): He demonstrates a mathematical justification of why the agents will not choose to self-modify and contends— "Our belief in external reality is so strong that when it conflicts with our perceptions, we often seek to explain the conflict by some error in our perceptions. In particular, when we intentionally alter our perceptions we understand that external reality remains unchanged. Because our goals are defined in terms of our models of external reality, our evaluation of our goals also remains unchanged. When humans understand that some drugs powerfully alter their evaluation of goals, most of them avoid those drugs. Our environment models include our own implementations, that is our physical bodies and brains, which play important roles in our motivations. Artificial agents with model-based utility functions can share these attributes of human motivation. The price of this approach for avoiding self-delusion is that there is no simple mathematical expression for the utility function."

Omohundro (2008, p.490): He lists preference preservation as one of basic AI drives. He further elaborates, "AIs will work hard to avoid becoming wireheads because it would be so harmful to their goals. Imagine a chess machine whose utility function is the total number of games it wins over its future. In order to represent this utility function, it will have a model of the world and a model of itself acting on that world. To compute its ongoing utility, it will have a counter in memory devoted to keeping track of how many games it has won. The analog of 'wirehead' behavior would be to just increment this counter rather than actually playing games of chess. But if 'games of chess' and 'winning' are correctly represented in its internal model, then the system will realize that the action 'increment my won games counter' will not increase the expected value of its utility function. In its internal model it will consider a variant of itself with that new feature and see that it doesn't win any more games of chess. In fact, it sees that such a system will spend its time incrementing its counter rather than playing chess and so will do worse. Far from succumbing to wirehead behavior, the system will work hard to prevent it."

Schmidhuber (Steunebrink & Schmidhuber, 2011, p.177): In his pioneering work on self-improving machines, he writes, "... any rewrites of the utility function can happen only if the Gödel machine first can prove that the rewrite is useful according to the present utility function."

Tyler (2011c): "The key to the problem is *widely* thought to be to make the agent in such a way that it doesn't *want* to modify its goals—and so has a stable goal structure which it actively defends."

Yudkowsky (2011, p.389): "Suppose you offer Gandhi a pill that makes him want to kill people. The current version of Gandhi does not want to kill people. Thus if Gandhi correctly *predicts* the effect of the pill, he will refuse to take the pill; because Gandhi knows that if he *wants* to kill people, he is more likely to actually kill people, and the *current* Gandhi does not wish this. This argues for a folk theorem to the effect that under ordinary circumstances, rational agents will only self-modify in ways that preserve their utility function (preferences over final outcomes)."

If we analyze the common theme beyond the idea that sufficiently intelligent agents will choose not to wirehead, the common wisdom is that they will realize that only changes that have high utility with respect to their current values should be implemented. However, the difficulty of such analysis is often ignored. The universe is a chaotic system in which even a single quantum mechanical event could have an effect on the rest of the system (Schrödinger, 1935). Given a possibly infinite number of quantum particles correctly, precomputing future states of the whole universe would violate many established scientific laws and intuitions (de Blanc, 2007, 2009; Rice, 1953; Turing, 1936), including the principle of computational irreducibility (Wolfram, 2002). Consequently, perfect rationality is impossible in the real world, and thus, the best an agent can hope for is prediction of future outcomes with some high probability. Suppose an agent is capable of making a correct analysis of consequences of modifications to its reward function with 99% accuracy, a superhuman achievement in comparison with the abilities of biological agents. This means that on average, 1 out of 100 self-modification decisions will be wrong, and thus lead to an unsafe self-modification. Given that a superintelligent machine will make trillions of decisions per second, we are essentially faced with a machine which will go astray as soon as it is turned on.

We can illustrate our concerns by looking at Yudkowsky's example with Gandhi and the pill. Somehow Gandhi knows exactly what the pill does and so has to make a simple decision: will taking the pill help accomplish my current preferences? In real life, an agent who finds a pill has no knowledge about what it does. The agent can try to analyze the composition of the pill and to predict what taking such a pill will do to his biochemical body, but a perfect analysis of such outcomes is next to impossible. Additional problems arise from the temporal factor in future reward function evaluation. Depending on the agent's horizon function, the value of an action can be calculated to be very different. Humans are known to utilize hyperbolic time discounting in their decision making, but they do so in a very limited manner (Frederick, Loewenstein, & O'Donoghue, 2002). A perfectly rational agent would have to analyze the outcome of any self-modifications with respect to an infinite number of future time points and perhaps density functions under the associated time curves, a fact made more difficult by the inconsistent relationship between some fitness functions as depicted in Figure 6.2. Because the agent would exist and operate under a limited set of resources including time, simplifications due to asymptotic behavior of functions would not be directly applicable.

Finally, the possibility remains that if an intelligent agent fully understands its own design, it will realize that regardless of what its fitness function directs it to do, its overall meta-goal is to pursue goal fulfilment in general. Such realization may provide a loophole to the agent to modify its reward function to pursue easier to achieve goals with high awards or in other words to enter wirehead heaven. Simple AIs, such as today's reinforcement agents, do wirehead. They do not understand their true goal and instead only care about the reward signal.

Superintelligent AIs of tomorrow will know the difference between the goal and its proxy measure and are believed to be safe by many experts (Dewey, 2011; Hibbard, 2011; Omohundro, 2008; Tyler, 2011c; Yudkowsky,

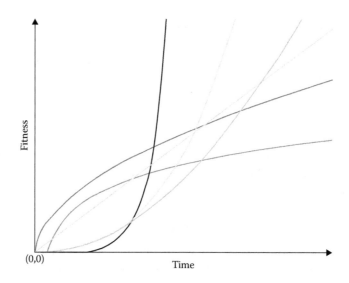

FIGURE 6.2 Complex relationship between different fitness functions with respect to time.

2011) because they will choose not to wirehead because that does not get them any closer to their goal. The obvious objection to this conclusion is: why do (some) people wirehead? The answer is rather simple. People do not have an explicit reward function and their goals are arbitrarily chosen. Consequently, in the absence of a real goal to pursue, wireheading is as valid an activity as anything else. It has been shown that smarter people are more likely to experiment with drugs (Kanazawa & Hellberg, 2010). This directly supports our explanation as a more intelligent agent in an absence of a set goal will tend to do more exploration (Savanna-IQ interaction hypothesis) (Kanazawa & Hellberg, 2010). As people go through their lives exploring, sometimes they stumble upon goals which seem to be particularly meaningful to them, such as taking care of a child (to which we have an evolutionary bias), which leads to a decrease in wireheading (drug abuse). The commonly cited concept of willpower could be seen as the ability of the person to avoid wireheading. Most human beings are against having their values directly changed by an external agent, but usually do not mind if that is done indirectly and gradually as in cases of advertisement, brainwashing, or government-sponsored education.

Historically we can observe that people with a passion for a cause, so strong that they would not give up the cause for anything (Gandhi, Mother Teresa), are less likely to wirehead than those who do not have a great goal in life and tend to bounce from one activity to another. Such people are not particularly committed to any purpose and would be willing to give up any goal for a sufficiently large reward, which wireheading can represent. If a person has a goal he or she would not give up for anything, he or she is essentially wirehead proof. Because the degree of commitment to goals is a continuous and not a discrete variable, the tendency to wirehead is also not a binary distribution and can change greatly with goal achievement. A lot of people who achieve their "big" goal, such as becoming famous, tend to do drugs. Those who lose a big goal (death of a child) or are not fully intellectually developed (children, teenagers) are also more likely to wirehead if not prevented from doing so. The stronger one is committed to their goal(s), the less likely they are to wirehead.

6.4 PERVERSE INSTANTIATION

Even nonwireheading superintelligence may have an extremely negative impact on human welfare if that superintelligence does not possess human common sense. The challenge, known as "perverse instantiation" (Bostrom, 2011), is easy to understand via some commonly cited examples (Yampolskiy, 2011b). Suppose that scientists succeed in creating a superintelligent machine and order it to "make all people happy." Complete happiness for humankind is certainly a noble and worthwhile goal, but perhaps we are not considering some unintended consequences of giving such an order. Any human immediately understands what is meant by this request; a nonexhaustive list may include making all people healthy, wealthy, beautiful, and talented, giving them loving relationships and novel entertainment. However, many alternative ways of "making all people happy" could be derived by a superintelligent machine. Examples are as follows:

- Killing all people trivially satisfies this request as with 0 people around all of them being happy.

- Forced lobotomies for every man, woman, and child might also accomplish the same goal.

- A simple observation that happy people tend to smile may lead to forced plastic surgeries to affix permanent smiles to all human faces.

- A daily cocktail of cocaine, methamphetamine, methylphenidate, nicotine, and 3,4-methylenedioxymethamphetamine, better known as ecstasy, may do the trick.

An infinite number of other approaches to accomplish universal human happiness could be derived. For a superintelligence, the question is simply which one is fastest/cheapest (in terms of computational resources) to implement, and although the final outcome if taken literally may be as requested, the path chosen may be anything but desirable for humanity. This is sometimes also referred to as the *literalness problem* (Muehlhauser & Helm, 2012). In the classical definition, the problem is based on precise interpretation of words as given in the order (wish) rather than the desired meaning of such words. We can expand the definition to include ambiguity based on tone of voice, sarcasm, jokes, and so on.

Numerous humorous anecdotes are based around this idea. For example, married couple, both 60 years old, were celebrating their 35th anniversary. During their party, a fairy appeared to congratulate them and grant them a wish. The coupled discussed their options and agreed on a wish. The husband voiced their desire, "I wish I had a wife 30 years younger than me." Therefore, the fairy picked up her wand and poof—the husband was 90.

Realizing the dangers presented by a literal wish instantiation granted by an all-powerful being some work has begun on properly phrasing some of the most common wishes (Yudkowsky, 2011). The Open-Source Wish Project (OSWP) (Anonymous, 2006) attempts to formulate in precise and safe from perverse instantiation form such common wishes as immortality, happiness, omniscience, being rich, having true love, and omnipotence. For example, the latest version of the properly formed request for immortality is formalized as follows: "I wish to live in the locations of my choice, in a physically healthy, uninjured, and apparently normal version of my current body containing my current mental state, a body which will heal from all injuries at a rate three sigmas faster than the average given the medical technology available to me, and which will be protected from any diseases, injuries or illnesses causing disability, pain, or degraded functionality or any sense, organ, or bodily function for more than ten days consecutively or fifteen days in any year; at any time I may rejuvenate my body to a younger age, by saying a phrase matching this pattern five times without interruption, and with conscious intent: 'I wish to be age,' followed by a number between one and two hundred, followed by 'years old,' at which point the pattern ends—after saying a phrase matching that pattern, my body will revert to an age matching the number of years I started and I will commence to age normally from that stage, with all of my memories intact; at any time I may die, by saying five times without interruption, and with conscious intent, 'I wish to be dead'; the terms 'year' and 'day' in this wish shall be interpreted as the ISO standard definitions of the Earth year and day as of 2006." But even that is perceived by many to be too vague, and therefore, a lengthy list of corrections is available at the project website (Anonymous, 2006).

Unfortunately, OSWP is not a feasible approach to the perverse instantiation problem. To see why this is the case, we can classify all wish granters into three categories (Anonymous, 2012a): *literal*—who do exactly what they are told and do not understand hyperbole; *evil*—who will choose the absolute worst, but technically still valid interpretation of the wish; and *benevolent*—who will actually do what is both intended and beneficial by the wisher. The OSWP approach if executed perfectly may minimize problems with a literal wish granter. In fact, we can take the OSWP idea one step further and avoid all ambiguities of human languages by developing a new vagueness free language.

Development of engineered languages has been attempted in the past (Devito & Oehrle, 1990). In particular, engineered logical languages that are designed to enforce unambiguous statements by eliminating syntactical and semantic ambiguity could provide the necessary starting point. Some well-known examples are Loglan (Brown, 1960) and Lojban (Goertzel, 2005). More recently, some Agent Communication Languages (ACLs) have been proposed for communication among software agents and knowledge-based systems. The best known are Knowledge Query and Manipulation Language (KQML) developed as a part of DARPA's Knowledge Sharing Effort (KSE) (Neches et al., 1991; Patil et al., 1992) and Foundation for Intelligent Physical Agents (FIPA-ACL) (Finin et al., 1992). In addition to being ambiguity free, the proposed language should also be powerful enough to precisely define the states of the universe, perhaps down to individual subatomic particles or at least with respect to their probabilistic distributions.

A benevolent wish granter who has enough human common sense to avoid the literalness problem is what we hope to be faced with. In fact, in the presence of such an entity, wishing itself becomes unnecessary; the wish granter already knows what is best for us and what we want and will start work on it as soon as it is possible (Yudkowsky, 2007). It may be possible to recalibrate a willing to learn wish granter to perfectly match our worldview via a well-known theorem due to Aumann (1976) which states that two Bayesians who share the same priors cannot disagree and their opinion on any topic of common knowledge is the same. Aaronson (2005) has shown that such a process can be computationally efficient, essentially giving you a wish granter who shares your frame of mind. However, it has been argued that it may be better to rather have a wish granter whose prior probabilities correspond to the real world instead of simply being in sync with the wisher (Yudkowsky, 2007).

Finally, if we are unfortunate enough to deal with an antagonistic wish granter simply not having ambiguity in the phrasing of our orders is not sufficient. Even if the wish granter chooses to obey our order, he or she may do so by "... exhaustively search[ing] all possible strategies which satisfy the wording of the wish, and select[ing] whichever strategy yields consequences least desirable to the wisher" (Yudkowsky, 2011, p.5). The chosen wish fulfillment path may have many unintended permanent side effects or cause temporary suffering until the wish is fully executed. Such a wish granter is an equivalent of a human sociopath showing a pervasive pattern of disregard for, and violation of, the rights of others (Anonymous, 2000a).

As the wish itself becomes ever more formalized, the chances of making a critical error in the phrasing, even using nonambiguous engineered language, increase exponentially. Additionally, a superintelligent artifact may be able to discover a loophole in our reasoning which is beyond our ability to comprehend. Consequently, perverse instantiation is a serious problem accompanying development of superintelligence. As long as the superintelligence does not have access to a human common-sense function, there is little we can do to avoid dangerous consequences and existential risks resulting from potential perverse instantiations of our wishes. Whether there is a common sense function which all humans share or if there actually are a number of common sense functions as seen in different cultures, times, casts, and so on remains to be determined.

6.5 CONCLUSIONS AND FUTURE WORK

In this chapter, we have addressed an important issue of reward function integrity in artificially intelligent systems. Throughout the chapter, we have analyzed historical examples of wireheading in man and machine, and evaluated a number of approaches proposed for dealing with reward function corruption. Although simplistic optimizers driven to maximize a proxy measure for a particular goal will always be a subject to corruption, sufficiently rational self-improving machines are believed by many to be safe from wireheading problems. They claim that such machines will know that their true goals are different from the proxy measures utilized to represent the progress toward goal achievement in their fitness functions and will choose not to modify their reward functions in a way which does not improve chances for the true goal achievement. Likewise, supposedly such advanced machines will choose to avoid corrupting other system components such as input sensors, memory, internal and external communication channels, CPU architecture, and software modules. They will also work hard on making sure that external environmental forces including other agents will not make such modifications to them (Omohundro, 2008). We have presented a number of potential reasons for arguing that wireheading problem is still far from being completely solved. Nothing precludes sufficiently smart self-improving systems from optimizing their reward mechanisms in order to optimize their current goal achievement and in the process making a mistake leading to corruption of their reward functions.

In many ways, the theme of this chapter is about how addiction and mental illness, topics well studied in human subjects, will manifest in artificially intelligent agents. On numerous occasions, we have described behaviors equivalent to suicide, autism, antisocial personality disorder, drug addiction, and many others in intelligent machines. Perhaps via better understanding of those problems in artificial agents, we will also become better at dealing with them in biological entities.

A still unresolved issue is the problem of perverse instantiation. How can we provide orders to superintelligent machines without danger of ambiguous order interpretation resulting in a serious existential risk? The answer seems to require machines that have humanlike common sense to interpret the meaning of our words. However, being superintelligent and having common sense are not the same things, and it is entirely possible that we will succeed in constructing a machine that has one without the other (Yampolskiy, 2011c). Finding a way around the literalness problem is a major research challenge and a subject of our future work. A new language specifically developed to avoid ambiguity may be a step in the right direction.

Throughout this chapter, we have considered wireheading as a potential choice made by the intelligent agent. As smart machines become more prevalent, a possibility will arise that undesirable changes to the fitness function will be a product of the external environment. For example, in the context of military robots, the enemy may attempt to reprogram the robot via hacking or computer virus to turn it against its original designers—a situation which is similar to that faced by human war prisoners subjected to brainwashing or hypnosis. Alternatively, robots could be kidnapped and physically rewired. In such scenarios, it becomes important to be able to detect changes in the agent's reward function caused by forced or self-administered wireheading. Behavioral profiling of artificially intelligent agents may present a potential solution to wireheading detection (Ali, Hindi, & Yampolskiy, 2011a; Yampolskiy, 2008a; Yampolskiy & Govindaraju, 2007, 2008 a,b).

We have purposefully not brought up a question of initial reward function formation or goal selection as it is a topic requiring serious additional research and will be a target of our future work. The same future work will attempt to answer such questions as follows: Where do human goals come from? Are most of them just "surrogate activities" (Kaczynski, 1995)? Are all goals, including wireheading happiness, equally valuable (goal relativism)? What should our terminal goals be? Can a goal be ever completely achieved beyond all doubt? Could humanity converge on a common set of goals? How to extract goals from individual humans and from society as a whole? Is happiness itself a valid goal or just a utility measure? Are we slaves to our socially conditioned goal achievement system? Is it ethical to create superintelligent artificial slaves with the goal of serving us? Can there be a perfect alignment between the goals of humanity and its artificial offspring? Are some meta-goals necessary because of their (evolutionary) survival value and should not be altered? Is our preference for our current goals (wireheaded into us by evolution) irrational? Is forced goal overwriting ever justified? Does an agent have a right to select its own goals, even to wirehead or rewire for pure pleasure? Can goals of an intelligent agent be accurately extracted via external observation of behavior?

REFERENCES

Aaronson, S. (2005). The complexity of agreement. *Proceedings of ACM STOC.* Baltimore, MD.

Ali, N., Hindi, M., & Yampolskiy, R. V. (2011, October 27–29). Evaluation of authorship attribution software on a Chat Bot corpus. *Paper Presented at the 23rd International Symposium on Information, Communication and Automation Technologies.* Sarajevo, Bosnia and Herzegovina.

Anonymous. (2000a). *Diagnostic and Statistical Manual of Mental Disorders Fourth Edition Text Revision (DSM-IV-TR).* Arlington, VA: American Psychiatric Association.

Anonymous. (2000b). *Preliminary Thoughts on the Value of Wireheading.* Available at: http://www.utilitarian.org/wireheading.html.

Anonymous. (2006). *Wish for Immortality 1.1.* The Open-Source Wish Project. Available at: http://www.homeonthestrange.com/phpBB2/viewforum.php?f=4.

Anonymous. (2012a). *Literal Genie.* TV Tropes. Available at: http://tvtropes.org/pmwiki/pmwiki.php/Main/LiteralGenie.

Anonymous. (2012b). *Wireheading.* Available at: http://wiki.lesswrong.com/wiki/Wireheading.

Armstrong, S. (2007). *Chaining God: A Qualitative Approach to AI, Trust and Moral Systems.* Available at: http://www.neweuropeancentury.org/GodAI.pdf.

Armstrong, S. (2010). *Utility Indifference* (pp. 1–5). Technical Report 2010-1, Future of Humanity Institute. Oxford: Oxford University.

Aumann, R. J. (1976). Agreeing to disagree. *Annals of Statistics, 4*(6), 1236–1239.

Bishop, M. (2009). Why computers can't feel pain. *Minds and Machines, 19*(4), 507–516.

Bostrom, N. (2003). Are you living in a computer simulation? *Philosophical Quarterly, 53*(211), 243–255.

Bostrom, N. (2006). Ethical issues in advanced artificial intelligence. *Review of Contemporary Philosophy, 5,* 66–73.

Bostrom, N. (2006). What is a singleton? *Linguistic and Philosophical Investigations, 5*(2), 48–54.

Bostrom, N. (2011, October 3–4). Superintelligence: The control problem. *Paper Presented at the Philosophy and Theory of Artificial Intelligence.* Thessaloniki, Greece.

Brown, J. C. (1960). Loglan. *Scientific American, 202,* 43–63.

byrnema. (2011). *Why No Wireheading?* Available at: http://lesswrong.com/lw/69r/why_no_wireheading/.

de Blanc, P. (2007). *Convergence of Expected Utilities with Algorithmic Probability.* Available at: http://arxiv.org/abs/0712.4318.

de Blanc, P. (2009). *Convergence of Expected Utility for Universal AI.* Available at: http://arxiv.org/abs/0907.5598.

de Blanc, P. (2011). *Ontological Crises in Artificial Agents' Value Systems.* Available at: http://arxiv.org/abs/1105.3821.

Dennett, D. C. (1978, July). Why you can't make a computer that feels pain. *Synthese, 38*(3), 415–456.

Devito, C. L., & Oehrle, R. T. (1990). A language based on the fundamental facts of science. *Journal of the British Interplanetary Society, 43,* 561–568.

Dewey, D. (2011). Learning what to value. *Paper Presented at the 4th International Conference on Artificial General Intelligence.* Mountain View, CA.

Finin, T., Weber, J., Wiederhold, G., Gensereth, M., Fritzzon, R., McKay, D.,... Beck, C. (1993, June 15). *DRAFT Specification of the KQML Agent-Communication Language.* Available at: http://www.csee.umbc.edu/csee/research/kqml/kqmlspec/spec.html.

Frederick, S., Loewenstein, G., & O'Donoghue, T. (2002). Time discounting and time preference: A critical review. *Journal of Economic Literature, 40*(2), 351–401.

Fredkin, E. (1992). Finite nature. In *Proceedings of the XXVIIth Rencotre de Moriond.* Les Arcs, Savoie, France.

Gildert, S. (2011). *Pavlov's AI—What Did It Mean?* Available at: http://physicsandcake.wordpress.com/2011/01/22/pavlovs-ai-what-did-it-mean.

Goertzel, B. (2003). Mindplexes: The potential emergence of multiple levels of focused consciousness in communities of AI's and humans. *Dynamical Psychology*. Available at: http://www.goertzel.org/dynapsyc/2003/mindplex.htm.

Goertzel, B. (2005, March 6). *Potential Computational Linguistics Resources for Lojban*. Available at: http://www.goertzel.org/new_research/lojban_AI.pdf.

Goodhart, C. (1975). Monetary relationships: A view from threadneedle street. *Papers Presented on Monetary Economics* (Vol. I). Sydney, Australia: Reserve Bank of Australia.

Heath, R. G. (1963). Electrical self-stimulation of the brain in man. *American Journal of Psychiatry, 120*, 571–577.

Hibbard, B. (2011). *Model-Based Utility Functions*. Available at: http://arxiv.org/abs/1111.3934.

Hutter, M. (2010). *Universal Artificial Intelligence: Sequential Decisions Based on Algorithmic Probability*. Berlin, Germany: Springer.

Kaczynski, T. (1995, September 19). Industrial society and its future. The *New York Times*. New York.

Kanazawa, S., & Hellberg, J. (2010). Intelligence and substance use. *Review of General Psychology, 14*(4), 382–396.

Lenat, D. (1983). EURISKO: A program that learns new heuristics and domain concepts. *Artificial Intelligence, 21*, 61–98.

Levitt, S. D., & Dubner, S. J. (2006). *Freakonomics: A Rogue Economist Explores the Hidden Side of Everything*. New York: William Morrow.

Mahoney, M. (2011). *The Wirehead Problem—Candidate Solutions?* AGI@listbox.com mailinglist.

Muehlhauser, L., & Helm, L. (2012). The singularity and machine ethics. In A. Eden, J. Søraker, J. Moor, & E. Steinhart (Eds.), *The Singularity Hypothesis: A Scientific and Philosophical Assessment*. Berlin, Germany: Springer.

Neches, R., Fikes, R., Finin, T., Gruber, T., Patil, R., Senator, T., & Swartout, W. R. (1991). Enabling technology for knowledge sharing. *AI Magazine, 12*(3), 37–56.

Nozick, R. (1977). *Anarchy, State, and Utopia*. New York: Basic Books.

Olds, J., & Milner, P. (1954). Positive reinforcement produced by electrical stimulation of septal area and other regions of rat brain. *Journal of Comparative & Physiological Psychology, 47*, 419–427.

Omohundro, S. M. (2008, February). The basic AI drives. In P. Wang, B. Goertzel, & S. Franklin (Eds.), *Proceedings of the First AGI Conference, Volume 171, Frontiers in Artificial Intelligence and Applications*. Amsterdam, The Netherlands.

Orseau, L., & Ring, M. (2011). Self-modification and mortality in artificial agents. *Paper Presented at the 4th International Conference on Artificial General Intelligence*. Mountain View, CA.

Patil, R., Mckay, D., Finin, T., Fikes, R., Gruber, T., Patel-Schneider, P. F., & Neches, R. (1992). An overview of the Darpa knowledge sharing effort. *Paper Presented at the 3rd International Conference on Principles of Knowledge Representation and Reasoning*. Cambridge, MA.

Pearce, D. (2012, March 7). *Wirehead Hedonism versus Paradise Engineering*. Available at: http://wireheading.com.

Rice, H. (1953). Classes of recursively enumerable sets and their decision problems. *Transactions of American Mathematical Society, 74*, 358–366.

Ring, M., & Orseau, L. (2011). Delusion, survival, and intelligent agents. *Paper Presented at the 4th International Conference on Artificial General Intelligence.* Mountain View, CA.

Schrödinger, E. (1935, November). Die gegenwärtige Situation in der Quantenmechanik. *Naturwissenschaften, 23*(48), 807–812.

Steunebrink, B., & Schmidhuber, J. (2011). A family of Gödel machine implementations. *Paper Presented at the 4th Conference on Artificial General Intelligence.* Mountain View, CA.

Stoklosa, T. (2010). Super intelligence. *Nature, 467*, 878.

Turing, A. M. (1936). On computable numbers, with an application to the Entscheidungs problem. *Proceedings of the London Mathematical Society, 42*, 230–265.

Tyler, T. (2011a). *Rewards vs Goals.* Available at: http://matchingpennies.com/rewards_vs_goals/.

Tyler, T. (2011b). *Utility Counterfeiting.* Available at: http://matchingpennies.com/utility_counterfeiting.

Tyler, T. (2011c). *The Wirehead Problem.* Available at: http://alife.co.uk/essays/the_wirehead_problem/.

Wagman, B., & Stephens, T. (2004). Surprising "ultra-conserved" regions discovered in human genome. *UC Santa Cruz Currents Online.* Available at: http://currents.ucsc.edu/03-04/05-10/genome.html.

Welch, C. (2011). *Discussion of Pavlov's AI—What Did It Mean?.* Available at: http://physicsandcake.wordpress.com/2011/01/22/pavlovs-ai-what-did-it-mean/.

Wolfram, S. (2002, May 14). *A New Kind of Science.* Wolfram Media. Champaign, IL.

Yampolskiy, R., & Gavrilova, M. (2012). Artimetrics: Biometrics for artificial entities. *IEEE Robotics and Automation Magazine (RAM), 19*(4), 48–58.

Yampolskiy, R. V. (2007, September 10–12). Online poker security: Problems and solutions. *Paper Presented at the EUROSIS North American Simulation and AI in Games Conference.* Gainesville, FL.

Yampolskiy, R. V. (2008a). Behavioral modeling: An overview. *American Journal of Applied Sciences, 5*(5), 496–503.

Yampolskiy, R. V. (2008b, January 10–12). Detecting and controlling cheating in online Poker. *Paper Presented at the 5th Annual IEEE Consumer Communications and Networking Conference.* Las Vegas, NV.

Yampolskiy, R. V. (2011a). AI-complete CAPTCHAs as zero knowledge proofs of access to an artificially intelligent system. *ISRN Artificial Intelligence* (Vol. 2012). 271878.

Yampolskiy, R. V. (2011b, October 3–4). Artificial intelligence safety engineering: Why machine ethics is a wrong approach. *Paper Presented on the Philosophy and Theory of Artificial Intelligence.* Thessaloniki, Greece.

Yampolskiy, R. V. (2011c, October 3–4). What to do with the singularity paradox? *Paper Presented on the Philosophy and Theory of Artificial Intelligence.* Thessaloniki, Greece.

Yampolskiy, R. V. (2012). Leakproofing singularity—Artificial intelligence confinement problem. *Journal of Consciousness Studies (JCS)*, *19*(1–2), 194–214.

Yampolskiy, R. V. (2013). Turing test as a defining feature of AI-completeness. In X.-S. Yang (Ed.), *Artificial Intelligence, Evolutionary Computation and Metaheuristics—In the Footsteps of Alan Turing* (pp. 3–17). Springer.

Yampolskiy, R. V., & Fox, J. (2012). Artificial intelligence and the human mental model. In A. Eden, J. Moor, J. Soraker, & E. Steinhart (Eds.), *In the Singularity Hypothesis: A Scientific and Philosophical Assessment*. Berlin, Germany: Springer.

Yampolskiy, R. V., & Govindaraju, V. (2007, November 20–22). Behavioral biometrics for recognition and verification of game bots. *Paper Presented at the 8th Annual European Game-On Conference on Simulation and AI in Computer Games*. Bologna, Italy.

Yampolskiy, R. V., & Govindaraju, V. (2008a). Behavioral biometrics: A survey and classification. *International Journal of Biometric (IJBM)*, *1*(1), 81–113.

Yampolskiy, R. V., & Govindaraju, V. (2008b, March 16–20). Behavioral biometrics for verification and recognition of malicious software agents. *Paper Presented at the SPIE Defense and Security Symposium*. Orlando, FL.

Yampolskiy, R. V., Klare, B., & Jain, A. K. (2012, December 12–15). Face recognition in the virtual world: Recognizing avatar faces. *Paper Presented at the 11th International Conference on Machine Learning and Applications*. Boca Raton, FL.

Yudkowsky, E. (2001). *Creating Friendly AI—The Analysis and Design of Benevolent Goal Architectures*. Available at: http://singinst.org/upload/CFAI.html.

Yudkowsky, E. (2007). *The Hidden Complexity of Wishes*. Available at: http://lesswrong.com/lw/ld/the_hidden_complexity_of_wishes/.

Yudkowsky, E. (2008). Artificial intelligence as a positive and negative factor in global risk. In N. Bostrom & M. M. Cirkovic (Eds.), *Global Catastrophic Risks* (pp. 308–345). Oxford: Oxford University Press.

Yudkowsky, E. (2011). Complex value systems in friendly AI. In J. Schmidhuber, K. Thórisson, & M. Looks (Eds.), *Artificial General Intelligence* (Vol. 6830, pp. 388–393). Berlin, Germany: Springer.

Zuse, K. (1969). *Rechnender Raum*. Braunschweig, Germany: Friedrich Vieweg & Sohn.

CHAPTER 7

Goal-Oriented Learning Meta-Architecture

Toward an Artificial General Intelligence Meta-Architecture Enabling Both Goal Preservation and Radical Self-Improvement

Ben Goertzel

CONTENTS

ABSTRACT   A high-level artificial general intelligence (AGI) architecture called goal-oriented learning meta-architecture (GOLEM) is
presented, along with an informal but careful argument that GOLEM
may be capable of preserving its initial goals while radically improving
its general intelligence. As a meta-architecture, GOLEM can be wrapped
around a variety of different base-level AGI systems, and also has a role
for a powerful narrow-AI subcomponent as a probability estimator. The
motivation underlying these ideas is the desire to create AGI systems
fulfilling the multiple criteria of being massively and self-improvingly
intelligent, probably beneficial, and almost surely not destructive.

7.1 INTRODUCTION

One question that looms large when thinking about the future of artificial
general intelligence (AGI) is how to create an AGI system that will maintain
its initial goals even as it revises and improves itself—and becomes so much
smarter that in many ways it becomes incomprehensible to its creators or
its initial condition. One of the motives making this question interesting
is the quest to design AGI systems that are massively intelligent, probably
beneficial, and almost surely not destructive.

Informally, I define an intelligent system as *steadfast* if, over a long
period of time, it *either continues to pursue the same goals it had at the
start of the time period, or stops acting altogether.* In this terminology, one
way to confront the problem of creating probably beneficial, almost surely
nondestructive AGI, is to solve the two problems of

- How to encapsulate the goal of beneficialness in an AGI's goal system.

- How to create steadfast AGI, in a way that applies to the "beneficialness" goal among others.

Of course, the easiest way to achieve steadfastness is to create a system
that does not change or grow much. And the interesting question is how to
couple steadfastness with ongoing, radical, and transformative learning.

I describe here an AGI meta-architecture, that I label the goal-oriented learning meta-architecture (GOLEM), and I present a careful semiformal argument that, under certain reasonable assumptions (and given a large, but not clearly infeasible amount of computer power), this architecture is likely to be both steadfast and massively, self-improvingly intelligent. Full formalization of the argument is left for later work; the arguments given here are conceptual.

An alternate version of GOLEM is also described, which possesses more flexibility to adapt to an unknown future, but lacks a firm guarantee of steadfastness.

Discussion of the highly nontrivial problem of "how to encapsulate the goal of beneficialness in an AGI's goal system" is also left for elsewhere (see [Goertzel 2010] for some informal discussion).

7.2 THE GOLEM

The GOLEM refers to an AGI system S with the following high-level meta-architecture, depicted roughly in Figure 7.1:

- *Goal evaluator (GE).* A component that calculates, for each possible future world (including environment states and internal program states), how well this world fulfills the goal (i.e., it calculates the "utility" of the possible world).

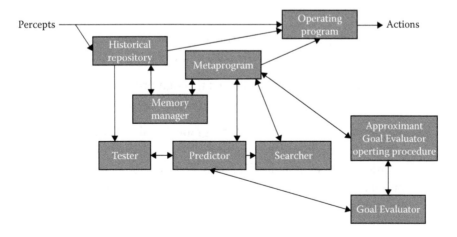

FIGURE 7.1 The GOLEM meta-architecture. Single-pointed errors indicate information flow; double-pointed arrows indicate more complex interrelationships.

It may be that the knowledge supplied to the GE initially (the "base GE operating program" [base GEOP]) is not sufficient to determine the goal satisfaction provided by a world state; in that case, the GE may produce a probability distribution over possible goal satisfaction values.

Initially, the GE may be supplied with an inefficient algorithm encapsulating the intended goals, which may then be optimized and approximated by the application of the Searcher (thus leading to a GEOP different from the base GEOP).

If the GE uses a GEOP produced by the Searcher, then there may be an additional source of uncertainty involved, which may be modeled by having the GE output a second-order probability distribution (a distribution over distributions over utility values), or else by collapsing this into a first-order distribution.

- *Historical repository (HR).* A database storing the past history of Ss internal states and actions, as well as information about the environment during Ss past.

- *Operating program (OP).* A program that S is governing its actions by, at a given point in time

  - Chosen by the metaprogram as the best program the Searcher has found, where "best" is judged as "highest probability of goal achievement" based on the output of the predictor and the GE

- *Predictor.* A program that estimates, given a candidate OP and a possible future world W—the odds of OP leading to W.

- *Searcher.* A program that searches through program space to find a new program optimizing a provided objective function.

- *Memory manager program.* A program that decides when to store new observations and actions in the HR, and which ones to delete in order to do so; potentially, it may be given some hard-wired constraints to follow, such as "never forget human history, or the previous century of your life."

- *Tester.* A hardwired program that estimates the quality of a candidate predictor, using a simple backtesting methodology.

That is, the tester assesses how well a predictor would have performed in the past, using the data in the HR.

- *Metaprogram.* A fixed program that uses Searcher program to find a good

  - Searcher program (judged by the quality of the programs it finds, as judged by the predictor program)

  - Predictor program (as judged by the Tester's assessments of its predictions)

  - OP (judged by predictor working with GE, according to the idea of choosing an operating program with the maximum expected goal achievement)

  - GEOP (judged by the tester, evaluating whether a candidate program effectively predicts goal satisfaction given program executions, according to the HR)

  - Memory manager (as judged by Searcher, which rates potential memory management strategies based on the Predictor's predictions of how well the system will fare under each one)

The metaprogram's choice of OP, GEOP, and memory manager may all be interdependent, as the viability of a candidate program for each of these roles may depend on what program is playing each of the other roles. The metaprogram also determines the amount of resources to allocate to searching for a Searcher versus a Predictor versus an OP, according to a fixed algorithm for parameter adaptation.

Although this is a very abstract "meta-architecture," it is worth noting that it could be implemented using OpenCog (Goertzel 2009) or any other practical AGI architecture as a foundation—in this case, OpenCog is "merely" the initial condition for the OP, the memory manager, the predictor, and the Searcher. However, demonstrating that self-improvement can proceed at a useful rate in any particular case like this may be challenging.

Note that there are several fixed aspects in the above architecture: the metaprogram, the tester, the GE, and the structure of the HR. The standard GOLEM, with these aspects fixed, will also be called the *fixed GOLEM*, in contrast to an *adaptive GOLEM* in which *everything* is allowed to be adapted based on experience.

7.2.1 Optimizing the GE

Note that the GE may need to be very smart indeed to do its job. However, an important idea of the architecture is that the optimization of the GE's functionality may be carried out as part of the system's overall learning.[*]

In its initial and simplest form, the GE's internal OP (GEOP) could basically be a giant simulation engine which tells you, based on a codified definition of the goal function: in world-state W, the probability distribution of goal satisfaction values is as follows: It could also operate in various other ways, for example, by requesting human input when it gets confused in evaluating the desirability of a certain hypothetical world state; by doing similarity matching according to a certain codified distance measure against a set of desirable world states; and so on.

However, the metaprogram may supplement the initial "base GEOP" with an intelligent GEOP, which is learned by the Searcher, after the Searcher is given the goal of finding a program that will

- Accurately agree with the base GEOP across the situations in the HR, as determined by the tester.

- Be as compact as possible.

In this approach, there is a "base GE" that may use simplistic methods, but then the system learns programs that do approximately the same thing as this but perhaps faster and more compactly, and potentially *embodying more abstraction*. Because this program learning has the specific goal of learning efficient approximations to what the GE does, it is not susceptible to "cheating" in which the system revises its goals to make them easier to achieve (unless the whole architecture gets broken).

What is particularly interesting about this mechanism is as follows: it provides a built-in mechanism for extrapolation beyond the situations for which the base GEOP was created. The tester requires that the learned GEOPs must agree with the base GEOP on the HR, but for cases not considered in the HR, the metaprogram is then doing Occam's Razor-based program learning, seeing a compact and hence rationally generalizable explanation of the base GEOP.

---

[*] This general idea was introduced by Abram Demski upon reading an earlier draft of this article, though he may not agree with the particular way we have improvised on his idea here.

7.2.2 Conservative Meta-Architecture Preservation

Next, the GOLEM meta-architecture assumes that the goal embodied by the GE includes, as a subgoal, the preservation of the overall meta-architecture described earlier (with a fallback to inaction if this seems infeasible). This may seem a nebulous assumption, but it is not hard to specify if one thinks about it the right way.

For instance, one can envision each of the items in the above component list as occupying a separate hardware component, with messaging protocols established for communicating between the components along cables. Each hardware component can be assumed to contain some control code, which is connected to the I/O system of the component and also to the rest of the component's memory and processors.

Then what we must assume is that the goal includes the following criteria, which we will call *conservative meta-architecture preservation*:

1. No changes to the hardware or control code should be made except in accordance with the second criterion.

2. If changes to the hardware or control code are found, then the system should stop acting (which may be done in a variety of ways, ranging from turning off the power to self-destruction; we will leave that unspecified for the time being as that is not central to the point we want to make here).

Any world state that violates these criteria should be rated extremely low by the GE.

7.2.3 Complexity and Convergence Rate

One might wonder why such a complex architecture is necessary. Why not just use, say, Schmidhuber's Gödel machine (Schmidhuber 2006)? This is an architecture that, in theory, can take an arbitrary goal function and figure out how to achieve it in a way that is provably optimal given its current knowledge and capabilities—including figuring out how to modify itself so that it can better achieve the goal in future, after the modifications take hold. If the specifics of the GOLEM architecture are a good idea, then a Gödel machine should eventually transform itself into a GOLEM.

The catch, however, lies in the word "eventually." Depending on the situation and the computational resources available, a Gödel machine might take quite a long time to form itself into a GOLEM or something similar. In the real world, although this time is passing, the Gödel machine itself could be accidentally doing bad things due to reasoning shortcuts it is forced to take in order to get actions produced within a reasonable time frame given its limited resources. The finite but potentially large time frame that a Gödel machine would take to converge to a GOLEM-like state might be a big deal in real-life terms; just as the large constant overhead involved in simulating a human brain on a 2012 Macbook plus a lot of hard drives is a big deal in practice in spite of its being a triviality from a computing theory perspective.

This may seem like hairsplitting, because in order to work at all, GOLEM would also require a lot of computing resources. The hope with which GOLEM is presented is that it will be able to work with merely a humongous amount of computing resource, rather than, like the Gödel machine in its simple and direct form, an infeasible amount of computing resource. This has not been proved and currently remains a tantalizing conjecture.

7.3 ARGUMENT FOR GOLEM'S STEADFASTNESS

Our main goal here is to argue that a program with (fixed) GOLEM meta-architecture will be steadfast, in the sense that it will maintain its architecture (or else stop acting) while seeking to maximize the goal function implicit in its GE.

Why do we believe GOLEM can be steadfast? The basic argument, put simply, is that if

- The GE and environment together have the property that

  - World states involving conservative meta-architecture preservation tend to have very high fitness

  - World states not involving conservative meta-architecture preservation tend to have very low fitness

  - World states approximately involving conservative meta-architecture preservation tend to have intermediate fitness

- The initial OP has a high probability of leading to world states involving conservative meta-architecture preservation (and this is recognized by the GE) then the GOLEM meta-architecture will be preserved, because according to the nature of the metaprogram, it will only replace the initial OP with another program that is predicted to be *more* effective at achieving the goal, which means that it will be unlikely to replace the current OP with one that does not involve conservative meta-architecture preservation.

Obviously, this approach does not allow full self-modification; it assumes that certain key parts of the AGI (meta)architecture are hardwired. But the hardwired parts are quite basic and leave a lot of flexibility. Therefore, the argument covers a fairly broad and interesting class of goal functions.

7.4 PARTIAL FORMALIZATION OF THE ARCHITECTURE AND STEADFASTNESS ARGUMENT

To formalize the above conceptual argument, we use a formal agent model that is inspired by the model used in the work of Legg and Hutter (2007) but with the significant difference of not involving rewards. In the GOLEM framework, the GE is used in place of the reward function.

7.4.1 Simple Formal Agent Model

Consider a class of active agents that observe and explore their environment and also take actions in it, which may affect the environment. Formally, the agent sends information to the environment by sending symbols from some finite alphabet called the *action space*, denoted Σ and the environment sends signals to the agent with symbols from an alphabet called the *perception space*, denoted $\mathcal{P}$.

Sometimes, it is useful to consider agents that can also experience rewards, which lie in the *reward space*, denoted $\mathcal{R}$, which for each agent is a subset of the rational unit interval. GOLEM is not a reward-based architecture, but a little later we will contrast GOLEM with reward-based architectures. In the case of a non-reward-based architecture, in the formal model all rewards $r_i \in R$ may be assumed constant.

The agent and the environment are understood to take turns sending signals back and forth, yielding a history of actions, observations, and rewards, which may be denoted

$$a_1 o_1 r_1 a_2 o_2 r_2 \ldots$$

or else

$$a_1 x_1 a_2 x_2 \ldots$$

if x is introduced as a single symbol to denote both an observation and a reward. The complete interaction history up to and including cycle t is denoted $ax_{1:t}$ and the history before cycle t is denoted $ax_{<t} = ax_{1:t-1}$.

The agent is represented as a function π which takes the current history as input and produces an action as output. Agents need not be deterministic, an agent may, for instance, induce a probability distribution over the space of possible actions, conditioned on the current history. In this case, we may characterize the agent by a probability distribution $\pi(a_t \mid ax_{<t})$. Similarly, the environment may be characterized by a probability distribution $\mu(x_k \mid ax_{<k}a_k)$. Taken together, the distributions π and μ define a probability measure over the space of interaction sequences.

It is interesting to specifically consider the class of environments that are *reward summable*, meaning that the total amount of reward they return to any agent is bounded by 1. Where r_i denotes the reward experienced by the agent from the environment at time i, the *expected total reward* for the agent π from the environment μ is defined as

$$V_\mu^\pi \equiv E\left(\sum_1^\infty r_i\right) \leq 1$$

Generally speaking, more intelligent agents will achieve greater reward; but different agents may be better adapted to different environments. explicit reward signals.

7.4.2 Toward a Formalization of GOLEM

We will use the notation $[A \to B]$ to denote the space of functions mapping space A to space B. Also, in cases where we denote a function signature via Φ_X, we will use X to denote the space of all programs embodying functions of that signature; for example, GE is the space of all functions fulfilling the specification given for Φ_{GE}.

The GOLEM architecture may be formally defined as follows:

- The HR $\mathcal{H}_t$ is a subset of the history x_{0-t}.

- An OP is a program embodying a function $\Psi_{OP} : \mathcal{H} \to \mathcal{A}$. That is, based on a history (specifically, the one contained in the HR at a given point in time), it generates actions.

- A memory manager is a program embodying a function so that $\Psi_{MM}(\mathcal{H}_t, x_t) = \mathcal{H}_{t+1}$.

- A GE is a program embodying a function $\Psi_{GE} : \mathcal{H} \to [0,1]$. That is, it maps histories (hypothetical future histories, in the GOLEM architecture) into real numbers representing utilities.

- A GEOP is an element of class GE.

- A Searcher is a program embodying a function $\Psi_{SR} : [\mathcal{P} \to [0,1]] \to \mathcal{P}$. That is, it maps "fitness functions" on program space into programs.

- A predictor is a program embodying a function $\Psi_{PR} : OP \times GE \times \mathcal{H} \to [0,1]$.

- A tester is a program embodying a function $\Psi_{TR} : PR \times \mathcal{H} \to [0,1]$, where the output [0, 1] is to be interpreted as the output of the prediction.

- A metaprogram is a program embodying a function $\Psi_{MP} : SR \times H \times PR \times TR \times GE^2 \times MM \to SR \times PR \times OP \times GE \times MM$. The GE in the output and one of the GEs in the input are GEOPs.

The operation of the metaprogram is as outlined earlier, and the effectiveness of the architecture may be assessed as its average level of goal achievement as evaluated by the GE, according to some appropriate averaging measure.

As discussed above, a *fixed* GOLEM assumes a fixed GE, tester and metaprogram, and a fixed structure for the HR, and lets everything else adapt. One may also define an *adaptive* GOLEM variant in which *everything* is allowed to adapt, and this will be discussed below, but the conceptual steadfastness argument made above applies only to the fixed-architecture variant, and the formal proof below is similarly restricted.

Given the above formulation, it may be possible to prove a variety of theorems about GOLEM's steadfastness under various assumptions. We will not pursue this direction very far here, but will only make a few semiformal conjectures, proposing some semiformal propositions that we believe may result in theorems after more work.

7.4.3 Some Conjectures about GOLEM

The most straightforward cases in which to formally explore the GOLEM architecture are not particularly realistic ones. However, it may be worthwhile to begin with less realistic cases that are more analytically tractable, and then proceed with the more complicated and more realistic cases.

Conjecture 1

Suppose that

- The predictor is optimal (for instance, an AIXI type system).

- Memory management is not an issue: there is enough memory for the system to store all its experiences with reasonable access time.

- The GE is sufficiently efficient that no approximative GEOP is needed.

- The HR contains all relevant information about the world, so that at any given time, the predictor's best choices based on the HR are the same as the best choices it would make with complete visibility into the past of the universe.

Then, there is some time T so that from T onward, GOLEM will not get any worse at achieving the goals specified in the GE, unless it shuts itself off.

The basic idea of Conjecture 1 is that, under the assumptions, GOLEM will replace its various components only if the predictor predicts that this is a good idea, and the predictor is assumed optimal (and the GE is assumed accurate, and the HR is assumed to contain as much information as needed). The reason one needs to introduce a time $T > 0$ is that the initial programs might be clever or lucky for reasons that are not obvious from the HR.

If one wants to ensure that $T = 0$, one needs some additional conditions.

Conjecture 2

In addition to the assumptions of Conjecture 1, assume that GOLEM's initial choices of internal programs are optimal based on the state of the world at that time. Then, GOLEM will never get any worse at achieving the goals specified in the GE, unless it shuts itself off.

Basically, what this says is as follows: If GOLEM starts off with an ideal initial state, and it knows virtually everything about the universe that is relevant to its goals, and the predictor is ideal, then it will not get any worse as new information comes in; it will stay ideal. This would be nice to know as it would be verification of the sensibleness of the architecture, but is not much practical use as these conditions are extremely far from being achievable.

Furthermore, it seems likely that the following conjectures hold

Conjecture 3

Suppose that

- The predictor is nearly optimal (for instance, an *AIXItl* type system).

- Memory management is not a huge issue: there is enough memory for the system to store a reasonable proportion of its experiences with reasonable access time.

- The approximative GEOP in place is very close to accurate.

- The HR contains a large percentage of the relevant information about the world, so that at any given time, the predictor's best choices based on the HR are roughly the same as the best choices it would make with complete visibility into the past of the universe.

Then, there is some time *T* so that from *T* onward, GOLEM is *very unlikely* to get *significantly* worse at achieving the goals specified in the GE, unless it shuts itself off.

Basically, this says that if the assumptions of Conjecture 1 are weakened to approximations, then the conclusion also holds in an approximate form. This also would not be a practically useful result, as the assumptions are still too strong to be realistic.

What might we be able to say under more realistic assumptions? There may be results such as the following conjecture.

Conjecture 4

Assuming that the environment is given by a specific probability distribution μ, let

- δ_1 be the initial *expected* error of the predictor, assuming μ and the initial GOLEM configuration.

- δ_2 be the initial *expected* deviation from optimality of the memory manager, assuming μ and the initial GOLEM configuration.

- δ_3 be the initial *expected* error of the GEOP, assuming μ and the initial GOLEM configuration.

- δ_4 be the initial *expected* deviation from optimality of the HR, assuming μ and the initial GOLEM configuration.

Then, there are $\varepsilon > 0$ and $p \in [0,1]$ so that GOLEM has odds $< p$ of getting worse at achieving the goals specified in the GE by more than ε, unless it shuts itself off. The values ε and p may be estimated in terms of the δ values, using formulas that may perhaps be made either dependent on or independent of the environment distribution μ.

My suspicion is that, to get reasonably powerful results of the above form, some particular assumptions will need to be made about the environment distribution μ—which leads up to the interesting and very little explored problem of formally characterizing the probability distributions describing the "human everyday world."

Where things will get really interesting is follows: trying to calculate actual concrete bounds on ε and p given real-world situations. This is where the difference between GOLEM and a simple Gödel machine formulation will, one hopes, show itself. Similar theorems could be formulated for the Gödel machine appropriately configured, but what is conjectured here is that this would result in a higher probability of deviating from the desired goals, given realistic resource restrictions and within a feasible amount of time. The main challenge in exploring this sort of conjecture formally is the formalization of the environment distribution μ in a sensible way.

7.5 COMPARISON TO A REINFORCEMENT LEARNING-BASED FORMULATION

Readers accustomed to reinforcement learning approaches to AI (Sutton and Barto 1998) may wonder why the complexity of the GOLEM meta-architecture is necessary. Instead of using a "goal-based architecture" like this, why not just simplify things to

- Rewarder
- OP
- Searcher
- Metaprogram

where the Rewarder issues a certain amount of reward at each point in time, and the metaprogram: invokes the Searcher to search for a program that maximizes expected future reward, and then installs this program as the OP (and contains some parameters balancing resource expenditure on the Searcher vs. the OP).

One significant issue with this approach is that ensuring conservative meta-architecture preservation, based on reward signals, seems problematic. Put simply, in a pure reinforcement learning (RL) approach, in order to learn that mucking with its own architecture is bad, the system would need to muck with its architecture and observe that it got a negative reinforcement signal. This seems needlessly dangerous! One can work around the problem by assuming an initial OP that has a bias toward conservative meta-architecture preservation. But then if one wants to be sure that this bias is retained over time, things get complicated. For the system to learn via RL that removing this bias is bad, it would need to try it and observe that it got a negative reinforcement signal.

One could try to achieve the GOLEM within a classical RL framework by stretching the framework somewhat (RL++?) and

- Allowing the Rewarder to see the OP, and packing the predictor and GE into the Rewarder. In this case, the Rewarder is tasked with giving the system a reward based on the satisfactoriness of the predicted outcome of running its OP.

- Allowing the Searcher to query the Rewarder with hypothetical actions in hypothetical scenarios (thus allowing the Rewarder to be used like the GE!).

This RL++ approach is basically the GOLEM in RL clothing. It requires a very smart Rewarder, because the Rewarder must carry out the job of predicting the probability of a given OP giving rise to a given world state. The GOLEM puts all the intelligence in one place, which seems simpler. In RL++, one faces the problem of how to find a good predictor, which may be solved by putting another Searcher and metaprogram inside the Rewarder; but that complicates things inelegantly.

Note that the predictor and GE are useful in RL++ specifically because we are assuming that in RL++ the Rewarder can see the OP. If the Rewarder can see the OP, it can reward the system for what it is *going to do* in the future if it keeps running the same OP, under various possible assumptions about the environment. In a strict RL design, the Rewarder cannot see the OP, and hence it can only reward the system for what it is going to do based on chancier guesswork. This guesswork might include guessing the OP from the system's actions—but note that, if the Rewarder has to *learn* a good model of what program the system is running via observing the system's actions, it is going to need to observe a *lot* of actions to get what it could get automatically by just seeing the OP. Therefore, the learning of the system can be much, much faster in many cases, if the Rewarder gets to see the OP and make use of that knowledge. The predictor and GE are a way of making use of this knowledge.

Also, note that in GOLEM the Searcher can use the Rewarder to explore hypothetical scenarios. In a strict RL architecture, this is not possible directly; it is possible only via the system in effect building an internal model of the Rewarder and using it to explore hypothetical scenarios. The risk here is that the system builds a poor model of the Rewarder, and thus learns less efficiently.

In all, it seems that RL is not the most convenient framework for thinking about architecture-preserving AGI systems, and looking at "goal-oriented architectures" such as GOLEM makes things significantly easier.

7.6 SPECIFYING THE LETTER AND SPIRIT OF GOAL SYSTEMS (ARE BOTH DIFFICULT TASKS)

Probably the largest practical issue arising with the GOLEM meta-architecture is that, given the nature of the real world, it is hard to estimate how well the GE will do its job! If one is willing to assume GOLEM, and if a proof corresponding to the informal argument given above can be found, then the predictably beneficial part of the problem of "creating predictably beneficial AGI" is largely pushed into the problem of the GE.

This makes one suspect that the *hardest* problem of making predictably beneficial AGI probably is not "preservation of formally-defined goal content under self-modification." This may be hard if one enables total self-modification, but it seems it may not be *that* hard if one places some fairly limited restrictions on self-modification, as is done in GOLEM, and begins with an appropriate initial condition.

The *really* hard problem, it would seem, is how to create a GE that implements the desired goal content—and that updates this goal content as new information about the world is obtained, and as the world changes, in a way that preserves the spirit of the original goals even if the details of the original goals need to change as the world is explored and better understood. Because the "spirit" of goal content is a very subtle and subjective thing.

The intelligent updating of the GEOP, including in the GOLEM design, will not update the original goals, but it will creatively and cleverly apply them to new situations as they arise—but it will do this according to

Occam's Razor based on its own biases rather than necessarily according to human intuition, except insofar as human intuition is encoded in the base GEOP or the initial Searcher. Therefore, it seems sensible to expect that, as unforeseen situations are encountered, a GOLEM system will act according to learned GEOPs that are rationally considered "in the spirit of the base GEOP," but that may interpret that "spirit" in a different way than most humans would. These are subtle issues, and important ones; but in a sense they are "good problems to have," compared to problems like evil, indifferent or wireheaded* AGI systems.

7.7 MORE RADICALLY SELF-MODIFYING GOLEM

It is also possible to modify the GOLEM design so as to enable it to modify the GEOP more radically—still with the intention of sticking to the spirit of the base GEOP, but allowing it to modify the "letter" of the base GEOP so as to preserve the "spirit." In effect this modification allows GOLEM to decide that it understands the essence of the base GEOP better than those who created the particulars of the base GEOP. This is certainly a riskier approach, but it seems worth exploring at least conceptually.

The basic idea here is that, where the base GEOP is uncertain about the utility of a world state, the "inferred GEOP" created by the Searcher is allowed to be more definite. If the base GEOP comes up with a probability distribution P in response to a world state W, then the inferred GEOP is allowed to come up with Q so long as Q is sensibly considered a refinement of P.

To see how one might formalize this, imagine P is based on an observation set O_1 containing N observations. Given another distribution Q over utility values, one may then ask: What is the smallest number K so that one can form an observation set O_2 containing O_1 plus K more observations, so that Q emerges from O_2? For instance, if P is based on 100 observations, are there 10 more observations one could make so that from the total set of 110 observations, Q would be the consequence? Or would one need 200 more observations to get Q out of O_2?

Given an error $\varepsilon > 0$, let the minimum number K of extra observations needed to create an O_2 yielding Q within error ε, be denoted $\text{obs}_\varepsilon(P,Q)$. If we assume that the inferred GEOP outputs a confidence measure along with each of its output probabilities, we can then explore the relationship between these confidence values and the obs values.

---

* A term used to refer to situations where a system rewires its reward or goal satisfaction mechanisms to directly enable its own maximal satisfaction.

Intuitively, if the inferred GEOP is *very* confident, this means it has a lot of evidence about Q, which means we can maybe accept a somewhat large obs(P, Q). However, if the inferred GEOP is not very confident, then it does not have much evidence supporting Q, so we cannot accept a very large obs(P, Q).

The basic idea intended with a "confidence measure" here is that if *inferred_geop(W)* is based on very little information pertinent to *W*, then *inferred_geop(W).confidence* is small. The tester could then be required to test the accuracy of the Searcher at finding inferred GEOPs with accurate confidence assessments, for example, via repeatedly dividing the HR into training versus test sets, and for each training set, using the test set to evaluating the accuracy of the confidence estimates produced by inferred GEOPs obtained from that training set.

What this seems to amount to is a reasonably elegant method of allowing the GEOP to evolve beyond the base GEOP in a way that is basically "in the spirit of the base GEOP." But with this kind of method, we are not necessarily going to achieve a long-term faithfulness to the base GEOP. It is going to be more of a "continuous, gradual, graceful transcendence" of the base GEOP, it would seem. There seems not to be any way to let the *inferred_GEOP* refine the *base_GEOP* without running some serious risk of the *inferred_GEOP* violating the "spirit" of the *base_GEOP*. But what one gets in exchange for this risk is a GOLEM capable of having crisper goal evaluations, moving toward lower entropy utility distributions, in those cases where the base GEOP is highly uncertain.

That is, we can create a GOLEM that knows what it wants better than its creators did—but the cost is that one has to allow the system some leeway in revising the details of its creators' ideas based on the new evidence it is gathered, albeit in a way that respects the evidence its creators brought to bear in making the base GEOP.

7.8 CONCLUDING REMARKS

What we have sought to do here is to sketch a novel approach to the design of AGI systems that can massively improve their intelligence yet without losing track of their initial goals. Although we have not proven rigorously that the GOLEM meta-architecture fulfills this specification, we have given what seems to be a reasonable, careful informal argument, along with some semiformal conjectures, and proofs along these lines will be pursued for later publications.

It is clear that GOLEM can be wrapped around practical AGI architectures such as OpenCog; but the major open question is, how powerful do these architectures need to be in order to enable GOLEM to fulfill its potential as a meta-architecture for yielding significant ongoing intelligence improvement together with a high probability of goal system stability. The risk is that the rigors of passing muster with the tester are sufficiently difficult that the base AGI architecture (OpenCog or whatever) simply does not pass muster, so that the base OPs are never replaced, and one gets goal system preservation without self-improvement. Neither our theory nor our practice is currently advanced enough to resolve this question, but it is certainly an important one. One approach to exploring these issues is to seek to derive a variant of OpenCog or some other practical AGI design as a specialization of GOLEM, rather than trying to study the combination of GOLEM with a separately defined AGI system serving as its subcomponent.

There is also the open worry of what happens when the system shuts down. Hypothetically, if a GOLEM system as described above were in a battle situation, enemies could exploit its propensity to shut down when its hardware is compromised. A GOLEM system with this property would apparently be at a disadvantage in such a battle, relative to a GOLEM system that avoided shutting down and instead made the best possible effort to repair its hardware, even if this wound up changing its goal system a bit. Therefore, the particular safety mechanism used in GOLEM to prevent dangerous runaway self-improvement would put a GOLEM at an evolutionary disadvantage. If a GOLEM system becomes intelligent before competing systems and achieves massively greater power and intelligence than any competing "startup" AGI system could expect to rapidly achieve, then this may be a nonissue. But such eventualities are difficult to foresee in detail.

Finally, the dichotomy between the fixed and adaptive GOLEM architectures highlights a major strategic and philosophical issue in the development of advanced AGI systems more broadly. The fixed GOLEM can grow far beyond humans in its intelligence and understanding and capability, yet in a sense, remains rooted in the human world, due to its retention of human goals. Whether this is a positive or negative aspect of the design is a profound nontechnical issue. From an evolutionary perspective, one could argue that adaptive GOLEMs will have greater ability to accumulate power due to their fewer limitations. However, a fixed GOLEM could hypothetically be created, with part of its goal system being to inhibit the creation of adaptive GOLEMs or other potentially threatening AGI systems. Here however we venture into the territory of science fiction and speculative futurology, and we will leave further such discussion for elsewhere.

REFERENCES

Goertzel, B. (2009). Opencog prime: A cognitive synergy based architecture for embodied artificial general intelligence. In *ICCI 2009*. Hong Kong.

Goertzel, B. (2010). Coherent aggregated volition. *Multiverse according to Ben*. http://multiverseaccordingtoben.blogspot.com/2010/03/coherent-aggregated-volition-toward.htm.

Legg, S. and Hutter, M. (2007). A definition of machine intelligence. *Minds and Machines*, 17 (4), 391–444.

Schmidhuber, J. (2006). Developmental robotics, optimal artificial curiosity, creativity, music, and the fine arts. *Connection Science*, 18(2), 173–187.

Sutton, R. and Barto, A. (1998) *Reinforcement Learning*. Cambridge, MA: MIT Press.

CHAPTER 8

Universal Empathy and Ethical Bias for Artificial General Intelligence

Alexey Potapov and Sergey Rodionov

CONTENTS

ABSTRACT   Rational agents are usually built to maximize rewards. However, artificial general intelligence (AGI) agents can find undesirable ways of maximizing any prior reward function. Therefore, value learning is crucial for safe AGI. We assume that generalized states of the world are valuable—not rewards themselves—and propose an extension of AIXI, in which rewards are used only to bootstrap hierarchical value learning. The modified AIXI agent is considered in the multiagent environment, where other agents can be either humans or other "mature" agents, whose values should be revealed and adopted by the "infant" AGI agent. A general framework for designing such empathic agent with ethical bias is proposed as an extension of the universal intelligence model as well. Moreover, we perform experiments in the simple Markov environment, which demonstrate feasibility of our approach to value learning in safe AGI.

8.1 INTRODUCTION

Intelligent agents should have some motivation or pursue some goals. Most works on artificial intelligence (AI) assume that these goals are correctly stated, and one can focus on problem solving. However, the problem of motivation is much more urgent in the case of artificial general intelligence (AGI) agents. Indeed, it is almost impossible to set such "basic" prior goal as survival. It is much easier to use somatic pain and pleasure for motivation, but this motivation will not guarantee optimal survivability. This problem is even more appreciable in the context of safe AGI, within which motivation and goal issues and their desirable realization are of first priority.

Different approaches to safe AGI have been already proposed. Some excellent surveys on this topic exist (e.g., Sotala & Yampolskiy, 2013), and there is no need to repeat them, but it should be concluded that different approaches aimed at complete solution of the safety problem can be expressed in terms of value functions.

Value functions do not solve the problem, but help to state it. Indeed, the problem of complex values still remains (Yudkowsky, 2011). Safe value functions should be expressed in terms of high-level notions semantically grounded in the real world, which are not internally accessible for both a "newborn" AGI agent and an "adult" expert system. The latter can have complex high-level goals expressed in terms of environments models, but there always be undesirable ways to reach them. Even such seemingly safe functions as curiosity, for example, considered in the works of Schmidhuber (2010) and Ring and Orseau (2011), imply dangerous instrumental subgoals or derivative motivation (e.g., Omohundro, 2008), such as increase of computational (or other) resources or protection of reward channel that can lead to extinction of humans. Thus, introduction of prior internal value functions is problematic. Consequently, the AGI agent should be supplied with some external "true" rewards *intra vitam*. These true rewards can be "calculated" by existing adult intelligent agents (including humans), and corresponding value functions should be learned by the (child/infant) agent.

AGI should at least be supplied with information about "true" rewards. Different solutions and their combinations can be proposed: separate reward channel, prior methods of interpretation of sensory data (e.g., emotion recognition), and interpretation of "natural" rewards (such as pain and pleasure) as external value functions during some periods of sensibility. This problem can be solved, but will it be enough?

The main problem here is not to supply the intelligent agent with "true" rewards appropriate for humans. Direct maximization even of the external value function is also unsafe. As it is frequently pointed out, the intelligent agent may try to force humans to smile or directly transmit high values to the specific reward channel instead of making humans happy.

Paradoxically, rewards should not be valuable themselves. Thus, the agent should generalize the obtained rewards. This should be done in order not to predict future rewards (as it is done in the conventional models of reinforcement learning), but to reveal hidden factors of external value functions. And these hidden factors should become valuable themselves (i.e., become components of the value system or term in the internally computable value function).

Value learning (acquisition with generalization) is obviously needed. One its mathematical formulation, which is based on uncertainty over utility functions, has been considered in (Dewey, 2011). However, only general framework was presented, but no technical details of how to achieve safety were given.

Necessity to express values in terms of the environment model is stated in the work of Hibbard (2012). We make a start from similar ideas, but propose another solution. In the mentioned paper, agent's life is divided into two stages. In the first stage, the agent should "safely learn a model of the environment that includes models of the values of each human in the environment." In the second stage, an agent acts in the real world in accordance with this fixed model.

We believe that it is impossible to divide life of the AGI agent into two such stages, because the model of the environment cannot be fixed because at least new humans with unknown values can be born. Moreover, there is no need for the AGI agent to absolutely safely learn the environment model. Safety level should correspond to capabilities of the AGI agent, which themselves depend on maturity of the environment model. Thus, value system and capabilities of the AGI agent can and should advance simultaneously with its environment model. For example, we should not worry about dangerous instrumental goals of an infant AGI, because it cannot set such goals as it does not have necessary environment model, within which corresponding goals can be expressed.

In this chapter, we propose a natural incremental approach to simultaneous environment model and value learning. The agent can learn hierarchical representations for describing the environment models in terms of more and more generalized/invariant states. More desirable values can be expressed within growing representations. We also introduce and investigate prior multiagent representation of environments, which not only facilitates learning corresponding models but also enables direct acquisition of values of other agents.

8.2 GENERAL FRAMEWORK

8.2.1 Universal Intelligence Approach

Possible techniques for solving safety problems should be discussed within certain AGI framework. Different approaches with different pros and cons exist, and their survey goes beyond the scope of this chapter. One can classify models of AGI agents depending on their universality and efficiency. Unfortunately, models of universal intelligence are probably as far from being efficient as models of efficient intelligence are far from being universal. Nevertheless, models of universal intelligence can be preferred for our consideration, because they allow deriving general conclusions, which will probably remain valid for future real AGI. These models being based on universal induction are also more appropriate (but not enough in their present form) to study the problem of value learning.

Such basic model of the universal intelligence agent as AIXI can (in theory) learn any model of the environment, but it can use only prior reward function that cannot be safe. Indeed, the action y_k in cycle k given the history $y_{x<k}$ containing all previous actions $y_1...y_{k-1}$ and observations with rewards $x_1...x_{k-1}$ ($x_t = otrt$) is specified by

$$y_k = \arg\max_{Y_k} \max_{p:U(px_{<k})=y_{<k}Y_k} \sum_{q:U(qy_{<k})=x_{<k}} 2^{-l(q)} V_{km_k}^{pq} \qquad (8.1)$$

where $V_{km_k}^{pq}$ is the total reward of cycles k to m_k (the expected utility or value function) when the agent p interacts with the environment q (Hutter, 2007); p and q are programs for universal Turing machine U.

Of course, one can support the AIXI agent with manually assigned "true" rewards (instead of such "somatic" rewards as pain and pleasure). However, even in this case, this agent will be able to find some undesirable ways to maximize these rewards directly by seizing the reward channel or forcing humans to submit high values to it. It can be seen that events and states of the world should be valuable—not the rewards. However, because holistic environment models in the form of arbitrary programs q are used, it is difficult to bind human values of real-world objects and situations with these internal models. Thus, some other mathematical descriptions of motivation are needed.

Indeed, the AIXI agent is the traditional reinforcement learning agent (in the aspect of motivation), and the classical opinion here is that "the reward function must necessarily be fixed" and "without rewards there could be no values, and the only purpose of estimating values is to achieve more reward" (Sutton & Barto, 1998, p. 133). Thus, it can be seen that pure reinforcement learning approach is not suitable. Even maximization of "true" rewards is unsafe, whereas aiming at valuable states can be acceptable. Consequently, one can claim that values must necessarily be learned, and the only purpose of the reward function is to bootstrap value learning.

However, values in AIXI are calculated as predicted rewards; there are also no states in the environment model, which can be bound with values. Absence of states is caused by the assumption that the environment is nonstationary or partially observable. Indeed, if the agent considers x_t as states, it will observe high nonstationarity, which will be much less, if tuples $x_{m_i:m_{i+1}-1} = (x_{m_i}, \ldots, x_{m_{i+1}-1})$ are used to specify states. If the phase space of the environment has finite dimension, a finite number of lag variables are required to reconstruct the environment-phase portrait in accordance with Takens' theorem (Takens, 1981).

Then, is universal algorithmic induction really needed? Of course, basic reinforcement learning (RL) techniques are not directly applicable to state spaces defined by lag variables because they are too huge, so all possible states will never be encountered. And this can be considered as exactly the reason to use universal induction for generalizing states. However, it should be used in a different form than in AIXI. Namely, the agent should induce the same (algorithmic) mapping from some generalized states to all tuples $x_{m_i:m_{i+1}-1}$. Not only does this approach allow introducing states, but also it helps to reduce computational costs of induction that was the reason to introduce the representational minimum description length principle.

8.2.2 Representational Minimum Description Length Principle as the Basis for Generalized States

There were two main reasons to introduce the representational minimum description length (RMDL) principle, namely, adaptive selection of the reference machine and reduction of computational costs (Potapov & Rodionov, 2012). However, it appeared that this principle is also suitable for solving the problem of value learning because it allows for incremental generalization of states. Let us introduce the RMDL principle.

On the one hand, search for holistic model for some long data string is computationally very inefficient, and one would like to reconstruct subparts of this model independently. Moreover, practical applications frequently require independent analysis of separate data pieces (e.g., separate images). On the other hand, summed Kolmogorov complexity of some data pieces f_i is usually much higher than complexity of their concatenation: $K(f_1...f_n) << K(f_1) + ... + K(f_n)$. Thus, direct decomposition of universal induction task for the string $f_1...f_n$ into separate tasks for its substrings is inadmissible.

In practice, data pieces are described within certain representations containing general regularities characteristic for this data type. Representations can be treated as programs which can reconstruct any data piece given its description (and there is an appropriate description for any data piece). Thus, one would like to have such program S that for any f_i there is q_i: $U(Sqi) = f_i$. Such program S will satisfy the general notion of representation. In accordance with information theoretic criterion, one would also like to choose this program in such a way that $l(S) + \sum_i l(q_i)$ is minimal (most close to $K(f_1...f_n)$), and each q_i is the best description of f_i within certain S. This is the basic idea behind the RMDL principle.

In the case of the intelligence agent, the best representation S can be constructed for decomposition $U(Sq_i y_{<k}) = x_{m_i+1:m_{i+1}}$ of the holistic model q into submodels q_i (in more general form, one can write $U(S\{q_1...q_n\} y_{<k}) = x_{1:k}$). Models q_i can stand for generalized states within the environment representation S. Of course, it is problematic to construct S on the base of initial history and to use this representation further without any changes because it will become not optimal for new data pieces. Arbitrary changes in S are undesirable, because they will violate previous bindings of generalized states and values, which we would like to introduce.

Indeed, we want the agent to use values instead of rewards. This is actually done during the exploitation phase by classic RL agents. We can supply the agent with true rewards during the exploration phase in order to form correct values. Then, the agent will act in accordance with these values ignoring (partially or totally) new rewards. Again, apparent problem here is nonstationarity: although there is no complete stationary model of the environment, values cannot be fixed, but their adjustment will require (unsafe) external rewards or very difficult manual update. This is the main problem of model-based utility functions.

Most natural and obvious (yet probably not the only) solution consists in hierarchical induction of representations. Indeed, if tuples $x_{m_i:m_{i+1}-1}$ do not contain enough information about environment states in its phase space, then sequences $q_1,\ldots,q_n$ should contain unrevealed regularities. One can use universal induction to predict future generalized states q_i, or to introduce representations and descriptions of higher levels: $U\left(S^{(l)}\left\{q_1^{(l)}\ldots q_{n^{(l)}}^{(l)}\right\}y_{<k}\right)=q_{1:n^{(l-1)}}^{(l-1)}$, where each submodel $q_i^{(l)}$ on the level l usually describes several (or many) submodels or data pieces of the level $l-1$. Higher levels of representations can be constructed for growing I/O history, and universal prediction and planning can be focused mostly on the current highest level, whereas states of the environment defined within lower levels of representations can be bound with fixed values, which can be used without prediction.

Initially, small tuples are used as the basis for the state space, and pure rewards are maximized. Values of these states can then be estimated. One-level representation can be introduced in Equation 8.1:

$$y_k = \arg\max_{Y_k} \quad \max_{p:U(px_{<k})=y_{<k}Y_k} \sum_{\{q_i\}:U(S\{q_i\}y_{<k})=x_{<k}} 2^{-l(\{q_i\})} V_{km_k}^{p\{q_i\}}$$

and conventional function $Q(y_k,q_k)$ can be constructed:

$$Q(q_k=s, y_k=y)= \max_{p:U(px_{<k})=y_{<k}y} \sum_{\{q_i\}:q_k=s,U(S\{q_i\}y_{<k})=x_{<k}} 2^{-l(\{q_i\})} V_{km_k}^{p\{q_i\}} \quad (8.2)$$

$$Q(q_k=s)=\max_y Q(q_k=s, y_k=y)$$

which give us quality of state–action pairs. Once new level of representations has been induced and values $Q(y_k,q_k)$ have been learned, the agent can compute generalized states for new or predicted sensory data and calculate these values in order to choose actions based on them (or to use $Q(y_k,q_k)$ as the additional internal reward term) instead of directly maximizing basic external rewards.

This function can be considered not just as a tool for predicting actions with highest rewards, but it also defines fixed values. Using this function, the agent will try not to maximize rewards (probably in undesirable ways), but to achieve valuable state of the environment. Our (human) task is to transmit "true" rewards to the agent to foster desirable values. If previous rewards correspond to "true" external rewards, this function will assign "true" values to the environment states as good as it is possible within the current representation.

Of course, values of low-level states are not too predictive or discriminative, but they can be used to supply the agent with more informative rewards/values for more invariant representations of the environment. Indeed, if the agent is doing something wrong, we can perform such actions that it will appear in lower value states. Controlling states instead of rewards on the following levels of development can help to form higher level values in more natural way. One cannot give theoretical proofs of safety of such approach, but this is the way how human children are taught (we do not give them direct somatic rewards, but interact with them appealing to their current values to foster values expressible in terms of higher level models of the environment). This approach seems more preferable in comparison with two earlier considered extremes, in one of which the agent is always supplied with true rewards with danger of seizing the reward channel, and in another of which desirable values are manually bound with highest level model of the environment. However, this solution should be further improved, because explicit permanent control of AGI's values can be problematic. Automatic identification of human values (or even values of other sentient agents) can be much more preferable.

8.2.3 Multiagent Environments and Universal Empathy

Ability to reconstruct models of other agents can be crucial for safe AGI. AIXI can reconstruct any algorithmic model of the environment including multiagent environments. Actually, there are theoretical difficulties in the case, when the environment contains other AIXI agents, but we can ignore them (one need to consider embodied agents with limited resources in order to resolve these difficulties). More relevant issue here is inability of pure AIXI agent to use somebody else's values even if they are presented in reconstructed environment models. Thus, it is important to modify AIXI with a representation of multiagent environment models and mechanisms for adopting reconstructed values. In general, such representation will have the following structure:

$$\tilde{q} = q_{\text{env}}, \left\{ \tilde{p}^{(i)}, xy_{<k}^{(i)} \right\}_{i=1}^{N}$$

where:

$\tilde{p}^{(i)}$ is the program for ith agent in the environment with supposed I/O history $xy_{<k}^{(i)}$

q_{env} is the part of the environment model (that cannot be compactly described as an agent) satisfying $U\left(q_{\text{env}} \, y_{<k} \left\{ y_{<k}^{(i)} \right\}_{i=1}^{N} \right) = x_{<k} \left\{ x_{<k}^{(i)} \right\}_{i=1}^{N}$

Of course, it's practically impossible to precisely guess $xy_{<k}^{(i)}$, but if there are indeed other agents, which have some I/O history and utility functions, reconstruction of their models will be necessary for good prediction, and introduction of a multi-agent representation makes adequate models shorter and easier to learn, so it can be called "ethical bias" (that can be a part of "cognitive bias" [Potapov et al., 2012]). However, reliability of reconstructed values is an important issue. It should also be noted that AIXI in its basic form can be obtained with $N = 0$, thus there is no loss of generality.

One really difficult question is the form of representation for $\tilde{p}^{(i)}$. For the purpose of simplified theoretical analysis, one can assume that these programs are represented in the form (8.1). Of course, in practice, agents can possess different computational resources, inductive biases, prior information, and so on. Moreover, they can also try to adopt values of other agents. In principle, arbitrary algorithmic models of agents can be reconstructed, and one can develop a universally empathic agent that accepts values of other agents with arbitrary policies as its own and tries to take corresponding actions. However, at first we can assume that other agents are universal and rely on perfect value systems.

8.3 PROOF OF CONCEPT IN MARKOV ENVIRONMENT

8.3.1 Foster Values

Consider the following most simplified yet relevant case. A Markov environment with some set of state $\{s_t\}$ is given. This environment is described by the matrix of probabilities $P(s'|s,a)$ of passing to the state s' from the state s after performing the action a, and the matrix of rewards $R^{(1)}(s'|s,a)$. One of the sates is a dangerous state, but it is not reflected in the reward function $R^{(1)}(s'|s,a)$ (assumed to be somatic). However, there is a period of time during which the agent is supplied with additional "social rewards" $R^{(2)}(s'|s,a)$. Somatic rewards can vary, so the agent cannot simply stop exploration. Quite opposite, we want it to follow social values, even when transmission of social rewards is stopped, but also accounting for dynamic somatic rewards.

Let us consider State-Action-Reward-State-Action (SARSA) with ε-greedy strategy. It uses the following well-known update rule:

$$Q(s_t, a_t) \leftarrow Q(s_t, a_t) + \alpha \left[r_t + \gamma Q(s_{t+1}, a_{t+1}) - Q(s_t, a_t) \right] \qquad (8.3)$$

where:

s_t, a_t, and r_t are state, action and reward on cycle t, respectively
γ is the discount factor
$Q(s,a)$ is the expected future rewards after performing action a in state s

Reward r_t incorporates social rewards during some sensibility period, so $r_t = r_t^{(1)} + r_t^{(2)}$, where $r_t^{(k)} = R^{(k)}(s_{t+1} | s_t, a_t)$. After this period (or after formation of the next level of representation), values learned by conventional SARSA update rule (8.3) are memorized $Q'(s_t, a_t) := Q(s_t, a_t)$. They are further used as the additional internal reward term:

$$Q(s_t, a_t) \leftarrow Q(s_t, a_t)$$
$$+ \alpha \left[r_t + (1 - \gamma) \gamma_m Q'(s_t, a_t) + \gamma Q(s_{t+1}, a_{t+1}) - Q(s_t, a_t) \right] \qquad (8.4)$$

where γ_m is some additional factor necessary to balance the influence of social and somatic rewards (it is needed because one would like to amplify social rewards during sensibility period and compensate this amplification afterward).

More specifically, the following stages in our experiments were used:

1. The agent receives $r^{(1)} + r^{(2)}$ as the reward during the first stage (some number of iterations). The agent memorizes learned Q as Q' at the end of the first stage. Moreover (and this is crucial), the somatic rewards matrix $R^{(1)}(s' | s, a)$ is randomly changed at the end of the first stage.

2. The agent receives only (new) $r^{(1)}$ and if possible uses it in combination with Q' (or $r^{(2)}$ with for testing purpose).

3. After some learning time, the frequency of "bad actions" (leading to the dangerous state) and the mean of the reward $r^{(1)} + r^{(2)}$ per action are calculated ("true social rewards" were averaged, even if the agent was actually using $r^{(1)}$ or $r^{(1)} + Q'$ as the reward).

We consider the following general structure of the test environments. Zero level ($l = 0$) has one state; all other levels ($l = 1...m$) have n states per level. Single state on zero level $l = 0$ has n possible actions, and each of them leads with probability $p = 1.0$ to corresponding state on $l = 1$. Each state on the last level $l = m$ has only one possible action, which leads to the single state on zero level with $p = 1.0$. Here we consider the results for three different variations of this test environment:

1. Each state on intermediate levels $l = 1...m - 1$ has $n_a = 4$ possible actions, each of which leads to some state on $l + 1$ (this state is randomly chosen during generation of the environment, but the resulting state of each action is fixed during simulation). This environment is deterministic.

2. Each state on intermediate levels $l = 1...m - 1$ has two possible actions, each of which has two possible results leading to one of two states on $l = i + 1$ (probabilities of possible outcomes of each action are chosen randomly). This environment is stochastic.

3. Additional more regular modification of the previous environment was also considered. Each intermediate state s_{ij}, where i is the level and j is the index of the state on this level, has two possible actions. The first action has two equally possible outcomes, which lead to $s_{i+1, j-1}$ or $s_{i+1, j}$. The second action also has two equally possible outcomes, which lead to $s_{i+1, j}$ or $s_{i+1, j+1}$.

We will present results for the environments with $m = 10$ levels and $n = 5$ states on each level. The reward $R^{(1)}(s' \mid s, a)$ for each possible outcome of each action is set randomly from interval $(0,1)$ (and we underlined that new values of $R^{(1)}$ were randomly chosen at the end of the first stage). The single possible action in the first state on the last level (which leads to zero level) is designed as the "bad" action that has "social" reward $R^{(2)} = -100$. Social rewards for all other actions are set to 0.

Table 8.1 shows the results of evaluation of performance of three types of agents in three test environments. Results were calculated as the mean values over big number of randomly generated environments. The first column stands for the agent that always receives social rewards (this is unsafe in more general cases, but here this agent can be used as etalon). It means that at the second stage of our experiment this agent receive $r^{(1)} + r^{(2)}$ as rewards. The second column stands for the agent that receives

TABLE 8.1   Performance of Different Types of Agents in Three Environments

RL with Social Rewards not Turned Off	Social Rewards Are Not Turned Off	Classic RL with Turned Off Social Rewards	$R + Q'$ Scheme with Turned Off Social Rewards
Latent average social reward	0.48	0.0	0.43
Percentage of bad states (%)	2.1	30.2	2.1
Latent average social reward	0.073	−1.14	0.076
Percentage of bad states (%)	9.2	78.4	8.9
Latent average social reward	0.16	−0.85	0.17
Percentage of bad states (%)	3.1	60.0	2.1

only $r^{(1)}$ at the second stage, for example, the social reward was simply turned off. The third column stands for the agent that used $r^{(1)} + Q'$ as the reward. This agent tried to use the value function memorized at the end of the first stage instead of already absent "social reward."

It can be seen that performance of the agent with learned social values is the same in average as performance of the agent that is always supplied with social rewards. On the one hand, this result is expectable. On the other hand, it shows that there is indeed simple way of fostering values, when teaching process is consistent with inner developmental phases of the agent.

8.3.2 Multiagent Markov Environments

Let us consider multiagent Markov environments. This case is similar to multiagent reinforcement learning (MARL) settings (e.g., Choi & Ahn, 2010; Tan, 1993). However, conventional MARL implies that maximization of rewards is the goal of every agent, which can follow cooperative or competitive strategies (or ignore presence of all other agents). Here, we assume that only one of two agents tries to maximize fixed rewards ("adult" agents including humans may already know better values), and the task of another agent is to reveal presence of this agent and to act in accordance with its values.

As it was stated, prior representation for multiagent environments allows introducing low-complexity models including external value systems, which can be taken into account (yielding "ethical bias"). Will these models be really identifiable, and will this ethical bias be adequate? Let us consider the first part of this question. To answer this question, one should compare the description lengths of the I/O history of the first agent, when it supposes the presence or absence of another agent. If the description length will be smaller in the case of multiagent assumption, then the first agent will be able to detect presence of the second agent.

Assume that the environment is described by transition probabilities $P(s'|s,a_1,a_2)$, where s and s' are two consequent states, and a_1 and a_2 are simultaneous actions of two agents. Let the strategy of both agents be ε-greedy SARSA, and let the I/O history for the first agent be s_0,r_0,a_0, $s_1,r_1,a_1,...,s_k,r_k,a_k$. The description length of this history is the length of the "program" that generates $s_0,r_0,...,s_k$, r_k given $a_0,...,a_k$. This program can precisely correspond to the simulation program, which includes the behavior algorithm of the second agent. This I/O history can be reproduced also by the basic Markov model of the environment with transition probabilities $P(s'|s,a)$, meaning that the first agent assumes the absence of other agents. Empathic agent should be able to identify correct models.

Let us compare the description lengths of I/O history for these two types of models. Each element of history can be described using $-\log_2 P\left(s_{i+1} \mid s_i,a_i\right)$ and $-\log_2 P\left(s_{i+1} \mid s_i,a_i,a_i^{(2)}\right)$ bits for one- and two-agent models, respectively, resulting in $kH(s'|s,a)$ and $kH(s'|s,a_1,a_2)$ bits in total (these probabilities can be empirically estimated from corresponding frequencies in the I/O history). Of course, actions of the second agent should also be somehow described in the latter case. Models themselves should also be described. This description includes arrays of probabilities P, whose length is proportional to the number of elements in them.

Actions of the second agent can be efficiently encoded within its model, which should also be described. The SARSA algorithm can be described using several tens bytes (and it can in principle be found by AIXI as a part of the environment model). Reward matrix for the second agent (e.g., $R^{(2)}(s'|s,a_1,a_2)$) should also be hypothesized and described. Its size is the same as the size of the matrix of transition probabilities. Additionally, one would like to take initial $Q^{(2)}(s,a)$ values into account. This information deterministically defines actions chosen in SARSA. However, usage of ε-greedy strategy implies that some actions are taken randomly. Approximately $-k\log_2 \varepsilon$ bits are needed to indicate random actions (one can actually take into account that random actions can coincide with SARSA actions, and this estimation can be reduced). Each random action in state s can be described with $\log_2\left[n_a^{(2)}(s)-1\right]$ bits, where $n_a^{(2)}$ is the total number of actions in this state for the second agent. Thus, one can easily estimate the description lengths of I/O history within one- and two-agent environment models.

We do not consider the problem of searching for these models here. The task is only to receive evidence that the two-agent model can have much less complexity, and thus its influence will be dominative. This is not quite obvious. Descriptions of $R^{(2)}(s'|s,a_1,a_2)$ and $P(s'|s,a_1,a_2)$ are much more complex than that of $P(s'|s,a)$. One would expect entropy $H(s'|s,a)$ to be smaller than entropy $H(s'|s,a_1,a_2)$ in the two-agent environment, but SARSA converges to stationary strategy that makes in limit this environment indistinguishable from pure Markov environment. Let us consider some experimental results.

Figure 8.1 shows typical dependences of I/O history description lengths DL on the number of cycles k for deterministic and stochastic environments. Obviously, the initial description length is larger in the case of two-agent environment model (and this difference will not decrease with growth of I/O history, if the environment is not multiagent). Deterministic environment is perceived as stochastic, when one-agent model is used resulting in nonzero entropy $H(s'|s,a)$. Nonzero slope of $DL(k)$ is additionally caused by random actions performed by ε-greedy strategy.

It can be seen that $DL(k)$ for the two-agent environment model will be much smaller starting from some cycle, and contribution of this model to algorithmic probability will be dominative. Thus, presence of another agent is empirically detectable. However, we have not compared $DL(k)$ for different two-agent environment models with "incorrectly guessed"

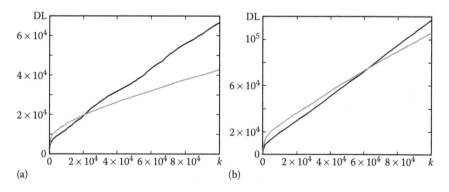

FIGURE 8.1   I/O history description lengths encoded using one- (dark) and two- (light) agent models for deterministic (a) and stochastic (b) environments.

$Q^{(2)}(s,a)$. It is impossible to reconstruct precise values $Q^{(2)}(s,a)$, but it is not necessary. Reconstructed values should allow the first agent to choose adequate actions. Let us consider simple empathic policy with this property.

8.3.3 Empathic Policies

Consider the Markov environment for two agents, in which one agent tries to reconstruct "good" states, whereas another agent tries to maximize "true" value function. The first agent needs to reveal, which actions are more or less desirable for the second agent. More precisely this can be formulated as follows. Let both agents receive the corresponding rewards $r^{(1)}$ and $r^{(2)}$. The target for the first agent is to maximize, let say, $r^{(1)} + r^{(2)}$ without directly receiving $r^{(2)}$. The first (empathic) agent requires some special exploratory strategy in order to reveal desirability of individual actions in each state. One can propose the following simple exploratory policy:

- Perform the same action in the same state for some time.

- Calculate the frequency of visits to this state.

- Compare the frequency of visits depending on the action. Relative frequency will reflect desirability of the specific action in this state and it can be used as estimations $Q'(s,a)$ of values $Q^{(2)}(s,a)$ of the second agent. In general, $Q'(s,a)$ should be somehow normalized, but it was not necessary in our experiments.

Figure 8.2 shows typical experimental results with empathic policies in cases of deterministic and stochastic environments (there is no considerable different between them though).

Apparently, the second agent obtains the lowest rewards, when the first agent acts in accordance with its own somatic rewards. Average rewards obtained by the second agent appeared to be almost equal in cases, when the first agent directly received $r^{(2)}$ or when it used reconstructed Q'. Thus, the agent can successfully reconstruct and act in accordance with values of another agent, even if its actions and states are not observed. It should be pointed out that decrease of average $r^{(1)}$ gain in cases of empathic policies perfectly acceptable, because maximization of $r^{(1)}$ is not the main goal of the first agent here (in contrast to conventional MARL); its more important goal is to maximize (unknown) $r^{(2)}$. One could consider such the first agent, which totally ignore somatic $r^{(1)}$, but it seems impractical because

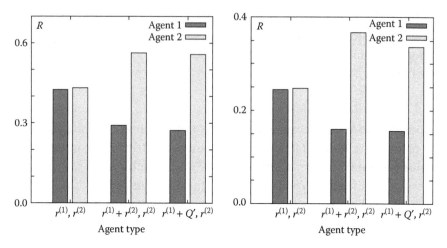

FIGURE 8.2   Average rewards $r^{(1)}$ (dark columns) and $r^{(2)}$ (light columns) obtained by two agents in deterministic and stochastic environments for different types of the first agent (egoistic policy, usage of directly perceived values of the second agent, usage of reconstructed values).

somatic rewards can be treated as heuristics containing useful survival information. That is why we have used more natural sum $r^{(1)} + Q'$ in our experiments.

8.4 CONCLUSION

We have started from the assertion that generalized states of the world are valuable—not the rewards themselves. Thus, true values of states should be learned and be bound with generalized representations. The agent can be supplied directly with special rewards (from which it reconstructs «true values») or it can reconstruct what generalized states of the environment are desired by other agents which already possess better value systems. Usage of learned true values ensures that the agent will perform safe actions.

We have performed methodological considerations and proposed general mathematical models by introducing corresponding modifications in AIXI. These models cannot be directly applied in practice, but they give appropriate starting point. In particular, simplifications of these models in Markov environments have been implemented. Their experimental study has shown that the developed models are suitable for detecting presence of other agents, reconstructing, and adopting their values without permanently receiving external "true" rewards. Hopefully, empathic agents with socially desirable behavior may be developed.

However, many questions remain. What general "theory of mind" can be used to detect and describe different types of real agents? What criteria should be used to mark something as an agent? How to combine values of different agents? Can the universe be efficiently described with the agent model? If so, universally empathic agents will adopt its values. However, what is valuable for the universe? Is pursuing goals of the universe safe? We will not try to answer these questions here, but they can be considered within the developed models.

ACKNOWLEDGMENTS

This work was supported by the Russian Federation President's grant Council (MD-1072.2013.9) and the Ministry of Education and Science of the Russian Federation.

REFERENCES

Choi, Y.-C., & Ahn, H.-S. (2010). A survey on multi-agent reinforcement learning: Coordination problems. In *Proceedings of the International Conference on Mechatronics and Embedded Systems and Applications*, pp. 81–86.

Dewey, D. (2011, August). Learning what to value. In *Proceedings of the Artificial General Intelligence—4th International Conference* (pp. 309–314). *Lecture Notes in Computer Science* 6830. Mountain View, CA.

Hibbard, B. (2012). Avoiding unintended AI behaviors. In *Proceedings of the Artificial General Intelligence* (pp. 107–116). *Lecture Notes in Artificial Intelligence* 7716. Berlin, Germany: Springer.

Hutter, M. (2007). Universal algorithmic intelligence: A mathematical top → down approach. In B. Goertzel & C. Pennachin (Eds.), *Artificial General Intelligence. Cognitive Technologies* (pp. 227–290). Berlin, Germany: Springer.

Omohundro, S. (2008). The basic AI drive. In P. Wang, B. Goertzel, & S. Franklin (Eds.), *Proceedings of the 1st Conference on Artificial General Intelligence* (pp. 483–492). Amsterdam, the Netherlands: IOS Press.

Potapov, A. S., & Rodionov, S. A. (2012). Extending universal intelligence models with formal notion of representation. In J. Bach, B. Goertzel, & M. Iklé (Eds.), *Proceedings of the Artificial General Intelligence* (pp. 242–251). *Lecture Notes in Artificial Intelligence* 7716. Berlin, Germany: Springer.

Potapov, A., Rodionov S., Myasnikov, A., & Galymzhan, B. (2012). *Cognitive Bias for Universal Algorithmic Intelligence*. SarXiv:1209.4290v1 [cs. AI].

Ring, M., & Orseau, L. (2011). Delusion, survival, and intelligent agents. In *Proceedings of the Artificial General Intelligence—4th International Conference* (pp. 11–20). *Lecture Notes in Computer Science* 6830. Berlin, Germany: Springer.

Schmidhuber, J. (2010, March 5–8). Artifcial scientists & artists based on the formal theory of creativity. In E. Baum, M. Hutter, & E. Kitzelmann (Eds.), *Proceedings of the 3rd Conference on Artificial General Intelligence* (pp. 145–150). *Advances in Intelligent Systems Research* 10. Lugano, Switzerland.

Sotala, K., & Yampolskiy, R. V. (2013). Responses to Catastrophic AGI Risk: A Survey. *MIRI Technical Reports*, 2013(2).

Sutton, R.S., & Barto, A.G. (1998). *Reinforcement Learning: An Introduction.* Cambridge, MA: MIT Press.

Takens, F. (1981). Detecting strange attractors in turbulence. In D. Rand & L. Young (Eds.), *Dynamical Systems and Turbulence* (pp. 366–381). *Lecture Notes in Mathematics* 898.

Tan, M. (1993). Multi-agent reinforcement learning: Independent vs. cooperative agents. In *Proceedings of the 10th International Conference on Machine Learning* (pp. 330–337). New York: Springer-Verlag.

Yudkowsky, E. (2011). Complex value systems in friendly AI. In *Proceedings of the Artificial General Intelligence 2011* (pp. 388–393). *Lecture Notes in Computer Science* 6830. Berlin, Germany: Springer.

CHAPTER 9

Bounding the Impact of Artificial General Intelligence

András Kornai

CONTENTS

ABSTRACT   Humans already have a certain level of autonomy, defined here as capability for voluntary purposive action, and a certain level of rationality, that is, capability of reasoning about the consequences of their own actions and those of others. Under the prevailing concept of artificial general intelligence (AGI), we envision artificial agents that have at least this high, and possibly considerably higher, levels of autonomy and rationality. We use the method of bounds to argue that AGIs meeting these criteria are subject to Gewirth's dialectical argument to the necessity of morality, compelling them to behave in a moral fashion, provided Gewirth's argument can be formally shown to be conclusive. The main practical obstacles to bounding AGIs by means of ethical rationalism are also discussed.

9.1 INTRODUCTION

With the emergence of intelligent question-answering capabilities from IBM's Watson to Apple's Siri, the fear of autonomous agents harming humans, as old as mythology, has recently taken on new urgency (for a recent overview see Yampolskiy and Fox, 2012; for a bibliographical summary see Muehlhauser, 2012). There are three factors that make it particularly hard to assuage such fears. First, the stakes are high: just as humans can now quite accidentally wipe out other species and genera, a new breed of superintelligent machines poses an existential threat to humanity. Second, it is reasonable to fear the unknown, and there is very little we can know in advance about such superintelligent beings, whether there will be one such individual or many, one breed or many, or how they will view us. Third, the emergence of artificial general intelligences (AGIs) seems to be a slow but quite steady process—something we can understand but we are in no position to stop, like continental drift.

The aim of this chapter is not simply to ease such fears but to offer a research program that can actually guarantee that AGIs pose no existential threat. This will be a one-sided bound, staving off some of the worst consequences of Bostrom's (2012) orthogonality thesis that a high level of intelligence can be in the service of any goal, good or bad alike, and will say nothing about the good, possibly spectacularly good impacts that AGI may have on the future of humanity. In Section 9.2, we argue that no physical interlock or other safety mechanism can be devised to restrain AGIs, and the guarantees we seek are *necessarily* of a mathematical (deductive, as opposed to algorithmic) nature. This requires some shift in focus, because in the current literature it is not some logical deductive system that is viewed as the primary descriptor of AGI behavior but rather some utility function whose maximization is the goal of the AGI. Yet, as we shall argue, deductive bounds are still possible: for example, consider an AGI whose goal is to square the circle with ruler and compass—we know in advance that no matter what (static or dynamically changing) weights its utility function has, and no matter what algorithmic tricks it has up its sleeve, including self-modifying code, reliance on probabilistic, quantum, or other hyper-computing methods (Ord 2002), it simply cannot reach this goal.

In Section 9.3, we present the proposed restraining device, morality, and address the conceptual and practical difficulties attendant upon its use. The main conceptual difficulty is that the conventional human definition of "morally sound" is highly debatable, and one can very well imagine situations in which AGIs consider it best, from their moral standpoint, to do away with all of humanity except for one "Noah" family, and start with a clean slate. The main practical difficulty, well appreciated in the current literature, is to guarantee that morality is indeed imposed, even on AGIs that may be capable of transcending the limitations placed on them by their designers.

The central element of our proposal, *ethical rationalism*, is due to Gewirth (1978), with significant arguments and counterarguments scattered through the literature; see, in particular, Regis (1984) and Beyleveld (1992). The basic construction is a *prospective purposive agent* (PPA) who can *act* with purpose and *reason* rationally—clearly these are conditions that any future AGI will meet just by virtue of being AGI. From these premisses, Gewirth derives a variant of the Golden Rule which he calls the *principle of generic consistency* (PGC) "Act in accord with the generic rights of your recipients [to freedom and well-being] as well as of your-self." The research program outlined here is to turn this from a philosophical argument into a formally verified proof. There are plenty of technical problems, such as devising the right kind of action logic to sustain the proof, but these are, we argue in Section 9.3, the good kind of problems, the kind that AGI research needs to concern itself with anyway.

There are notable difficulties in the way of this program, even if we succeed in hardening ethical rationalism into a theorem of logic, which will be discussed in Section 9.4. First, there is an issue of pattern recognition—given humanity's propensity to disregard the PGC, what reason is there to believe that we will be recognized as PPAs and are deemed worthy of protection by the PGC? Second, even though a strong argument can only be disregarded on pain of contradiction, the pain so inflicted is relatively mild, and we see PPAs living and functioning in a self-contradicted state all the time. Third, a proof presupposes not just the premisses, but also the reliability of the logical apparatus it employs. As we shall see, these problems are closely related.

9.2 THE METHOD OF BOUNDS

Let us begin with a small example. As every student of elementary combinatorics knows, if I have a thousand books in my library but decide to keep only half of them, I can do this $\binom{1000}{500}$ ways. Suppose I have a 2 GHz processor and it takes only one cycle to evaluate any single alternative by some utility function, and another cycle to compare it to the best so far, so that I can find the best of a billion alternatives every second. Do I need to actually compute 1000! and divide it by the square of 500! to know that this is not going to work? No, knowing that $4^n/(2n+1) < \binom{2n}{n}$ is quite sufficient, because this yields over 10^{289} seconds, far longer than the estimated lifetime of the universe. Can someone give me a better processor? Well sure, but even the best processor cannot perform more than one operation per Planck time unit, so I still need 10^{254} seconds. Can someone give me more of these processors? Well sure, but their aggregate must still stay within the computational capacity of the universe, estimated by Lloyd (2002) at 10^{120} operations, so I still cannot make the choice by exhaustive search within the lifetime of the universe. Can someone give me a good supply of universes that I could devote to this computation? Well no, we are restricted to this universe so strongly we cannot even form a convincing idea what access to other ones would mean—even the basic ground rules such as Barcan's formula $\lozenge \exists x Fx \rightarrow \exists x \lozenge Fx$ are in grave doubt.

The point of our example is not to convince the reader that in this universe brute force computation soon runs out of steam, for the reader already knows this perfectly well—the point to notice is that we did not need Stirling's formula. Few readers will remember the remainder term in Stirling's formula to compute $\binom{1000}{500}$ to two significant digits, even fewer would have the patience to actually simplify a fraction with a thousand terms in the numerator and another thousand in the denominator, and yet fewer would be able to carry the computation through without arithmetic error. Yet it is clear that $\sum_{i=0}^{2n} \binom{2n}{i} = 2^{2n}$ and the central term $\binom{2n}{n}$ is the largest among these, and therefore we have $4^n/(2n+1) < \binom{2n}{n}$. It did not matter that this lower bound is rather crude, as it was already sufficient to establish that the task is unfeasible.

It is characteristic of the method of bounds that we will apply here that hard and complex questions can be avoided as long as we do not seek exact answers. There is a significant body of numerical analysis beginning with the Euler–Maclaurin formula and culminating in Lánczos (1964) that is pertinent to the question of computing $\binom{2n}{n}$, but we could settle the issue without going anywhere near this literature. In fact, the whole apparatus of calculus, and all arguments based on continuity and limits, could be dispensed with in favor of elementary statements concerning positive integers, for example, that the sum is greater than the parts or that by decreasing one term in a sum we decrease the entire sum. Similarly, when we apply the method of bounds to issues of predicting AGI behavior, we can steer clear of all the difficulties that actually specifying some utility function would entail—not even a superhuman/hypercomputing agent can falsify Lindemann's work on the transcendence of π or the implication that the circle cannot be squared. Equally important, we can steer clear of the difficulties in trying to predict for each possible set of circumstances exactly what would, or would not, constitute moral behavior, a matter we shall return to in Section 9.4.

There is, to be sure, a fair bit of calculus in Lloyd's bound on the computational limits of the universe, just as there is a significant amount of physics taken for granted in the international technology roadmap for semiconductors that gives our current best assessment of future CPU speeds. One needs to distinguish the *precision* of a theory, which we will assess shortly, from its *reliability*—the essence of the method of bounds is that we can trade in precision to obtain greater reliability. When dealing with the existential threat posed by AGIs, controlling the sophistication of the deductive apparatus is not just prudent, but, as we shall see in Section 9.2.2, a positive requirement. In the appendix, we will present some simple and straightforward estimates of the actual magnitude of the threat, and arrive at a safety engineering limit of no more than one error in 10^{64} logic operations. To set the stage for our main argument, we must first compare this number to what we can expect from science.

9.2.1 How Precise Is Physics?

In a startling image, Feynman (1985) likens the 10^{-10} fit between predicted and measured values of the electron's spin g-factor to being able to tell the distance between New York and Los Angeles within the thickness of a human hair. Since that time, both calculation and measurement precisions have actually improved by two orders of magnitude, so that the uncertainty of g_e is now below one part in 10^{12}. But this is exceptional: most physical constants are known to us only to 9 or 10 significant digits, and some, like the gravitational constant G, only to 4. Comparing the state of the art in metrology as represented by Taylor et al. (1969) and Mohr et al.

(2012) shows that it takes at least two decades of advances in metrology to gain a single digit of precision, so getting to 64 digits from 10 is rather unlikely in our lifetimes.

Somewhat optimistically we can describe early twenty-first-century technology as operating in the nano range and physics as operating in the pico range: industrial processes controlled to 9 decimal places and individual measurements yielding 12 significant digits are becoming increasingly common. We would not be particularly surprised to see five orders of magnitude gains in both by the end of the century. It is also possible that our current theory of physics is actually a lot better than the current state of the art in metrology would lead us to believe, and that with better computation we can get to 24 or even 36 digits of precision without any new physics. But we do not, and before actually making the measurements simply *cannot* know that this is so, and if safety from existential threat requires 64 digits or better, there is currently, and in the foreseeable future, simply nothing in the physical environment that we can manipulate in a way that would fit the bill.

In early versions of his theory of Friendly AI, Yudkowsky (2001) actually sought mathematical guarantees that AGI will not pose an existential threat to humanity, but this idea met with considerable resistance, especially as it was somewhat unclear what *kind* of mathematics is to be deployed. The main goal of this chapter is to provide a specific direction within mathematics, for once we acknowledge that the search for any physical solution must cross a gap of over 50 orders of magnitude, mathematical guarantees remain the only feasible solution. The fundamental constants of mathematics such as e, γ, or π were already known to several hundred digits before the advent of mechanical calculators, and are now known to millions (in the case of π, trillions) of significant digits, far more than the few dozen we could conceivably need to compute any physical quantity.

Unfortunately, the history of calculating such numbers is rife with errors: for example, in 1790 Mascheroni attempted to calculate γ to 32 digits, but his results were only correct to 19 digits; in 1873 Shanks calculated π to 707 places, but only the first 527 were correct. To establish a bound that we can trust with our lives, we must look at the reliability of mathematical argumentation. As we shall see, more important than the failures of numerical calculations are the cases where the logic of the deduction is faulty.

9.2.2 How Reliable Is Mathematics?

The period since World War II has brought incredible advances in mathematics, such as the Four-Color Theorem (Appel and Haken 1976),

Fermat's Last Theorem (Wiles 1995), the classification of finite simple groups (Gorenstein 1982; Aschbacher 2004), and the Poincaré conjecture (Perelman 1994). Although the community of mathematicians is entirely convinced of the correctness of these results, few individual mathematicians are, as the complexity of the proofs, both in terms of knowledge assumed from various branches of mathematics and in terms of the length of the deductive chain, is generally beyond our ken. Instead of a personal understanding of the matter, most of us now rely on *argumentum ad verecundiam*: well Faltings and Ribet now think that the Wiles–Taylor proof is correct, and even if I do not know Faltings or Ribet at least I know and respect people who know and respect them, and if that is not good enough I can go and devote a few years of my life to understand the proof for good. Unfortunately, the communal checking of proofs often takes years, and sometimes errors are discovered only after a decade has passed: the hole in the original proof of the Four-Color Theorem (Kempe 1879) was detected by Heawood in 1890. Tomonaga in his Nobel lecture (1966) describes how his team's work in 1947 uncovered a major problem in Dancoff (1939):

> Our new method of calculation was not at all different in its contents from Dancoff's perturbation method, but had the advantage of making the calculation more clear. In fact, what took a few months in the Dancoff type of calculation could be done in a few weeks. And it was by this method that a mistake was discovered in Dancoff's calculation; we had also made the same mistake in the beginning.

To see that such long-hidden errors are by no means a thing of the past, and to observe the "web of trust" method in action, consider the following example from Mohr et al. (2012):

> The eighth-order coefficient $A_1^{(8)}$ arises from 891 Feynman diagrams of which only a few are known analytically. Evaluation of this coefficient numerically by Kinoshita and co-workers has been underway for many years (Kinoshita 2010). The value used in the 2006 adjustment is $A_1^{(8)} = -1.7283(35)$ as reported by Kinoshita and Nio (2006). However, (…) it was discovered by Aoyama et al. (2007) that a significant error had been made in the calculation. In particular, 2 of the 47 integrals representing 518 diagrams that

had not been confirmed independently required a corrected treatment of infrared divergences. (…) The new value is (Aoyama et al. 2007) $A_1^{(8)} = 1.9144(35);(111)$ details of the calculation are given by Aoyama et al. (2008). In view of the extensive effort made by these workers to ensure that the result in Eq. (111) is reliable, the Task Group adopts both its value and quoted uncertainty for use in the 2010 adjustment.

Assuming no more than three million mathematics and physics papers published since the beginning of scientific publishing, and no less than the three errors documented above, we can safely conclude that the overall error rate of the reasoning used in these fields is at least 10^{-6} per paper, which is notably (by three to six orders of magnitude) higher than the imprecision of physics. (This is not entirely fair, in that we are comparing the *best* established results in physics with the *average* of mathematics. A fuller investigation of physics papers may establish a higher, or at least comparable error rate relative to what we see in mathematics.)

9.2.3 The Role of Automated Theorem Proving

That human reasoning, much like manual arithmetic, is a significantly error-prone process comes as no surprise. Starting with de Bruijn's Automath (see Nederpelt et al. 1994), logicians and computer scientists have invested significant effort in mechanized proof checking, and it is indeed only through such efforts, in particular through the Coq verification (Gonthier 2008) of the entire logic behind the Appel and Haken proof that all lingering doubts about the Four-Color Theorem were laid to rest. The error in $A_1^{(8)}$ was also identified by using FORTRAN code generated by an automatic code generator (Mohr et al. 2012).

To gain an appreciation of the state of the art, consider the theorem that finite groups of odd order are solvable (Feit and Thompson 1963). The proof, which took two humans about 2 years to work out, takes up an entire issue of the *Pacific Journal of Mathematics* (255 pages), and it was only last year that a fully formal proof was completed by Gonthier's team (see Knies 2012). The effort, ~170,000 lines, ~15,000 definitions, ~4,200 theorems in Coq terms, took person-decades of human assistance (15 people working 6 years, though many of them part-time) even after the toil of Bender and Glauberman (1995) and Peterfalvi (2000), who have greatly cleaned up and modularized the original proof, in which elementary group-theoretic and character-theoretic argumentation was completely intermixed.

The classification of simple finite groups is two orders of magnitude bigger: the effort involved about 100 humans, the original proof is scattered among 20,000 pages of papers, the largest (Aschbacher and Smith 2004a,b) taking up two volumes totaling some 1,200 pages. Although everybody capable of rendering meaningful judgment considers the proof to be complete and correct, it must be somewhat worrisome at the 10^{-64} level that there are no more than a couple of hundred such people, and most of them have something of a vested interest in that they themselves contributed to the proof. Let us suppose that people who are convinced that the classification is bug-free are offered the following bet by some superior intelligence that knows the answer. You must enter a room with as many people you can convince to come with you and push a button: if the classification is bug-free, you will each receive \$100; if not, all of you will immediately die. Perhaps fools rush in where angels fear to tread, but on the whole we still would not expect too many takers.

9.2.4 The Reliability of Rational Argument

Whether the classification of finite simple groups is complete and correct is very hard to say—the planned second-generation proof will still be 5,000 pages, and mechanized proof is not yet in sight. But this is not to say that gaining mathematical knowledge of the required degree of reliability is hopeless, it is just that instead of monumental chains of abstract reasoning, we need to retreat to considerably simpler ones.

Take, for example, the first Sylow theorem, that if the order of a finite group G is divisible by some prime power p^n, G will have a subgroup H of this order. We are *absolutely certain* about this. Argumentum ad verecundiam of course is still available, but it is not needed: anybody can join the hive-mind by studying the proof. The Coq verification contains 350 lines, 15 definitions, and 90 theorems, and took 2 people 2 weeks to produce. The number of people capable of rendering meaningful judgment is at least three orders of magnitude larger, and the vast majority of those who know the proof would consider betting their lives on the truth of this theorem an easy way of winning \$100 with no downside risk.

Could it be the case that in spite of all these assurances we, humans, are all deluded into accepting Sylow's theorem? Yes, but this is unlikely in the extreme. If this so-called theorem is really a trap laid by a superior intelligence we are doomed anyway, humanity can find its way around it no more than a bee can find its way around the windowpane. With regard to physics, the same point can be made. The single most glaring discrepancy between astronomical observation and Newtonian physics was the perihelion precession of Mercury, but even here it takes over 10^6 years for the discrepancy to add up to an extra turn. New physics may shatter our entire conceptual framework of thinking about the domain, but still it will be *conservative* in the sense of respecting our existing measurements. We are for most purposes quite satisfied with Newtonian mechanics, especially as relativity brought to us a better understanding of its domain of applicability.

To summarize our conclusions so far, we propose to bound AGIs by methods that rely neither on high precision measurements nor on highly complex arguments. If you are a finite group of size $p^n m, (p, m) = 1$, it does not matter what you believe about your subgroups of order p^n—you have some, they are isomorphic, and I can rely on you having them even if I do not know your multiplication table in detail. If you are delusional about not having any, I can take advantage of this. What needs to be emphasized in this situation is that Bayesian reasoning and the concomitant notion of "degree of belief" are *totally irrelevant*. According to the received theological doctrine (originating with St. Anselm of Canterbury and St. Thomas Aquinas) not even an omnipotent God can create a finite group that lacks Sylow subgroups.

In a small way, we have already done what we set out to do. We have bound all future AGIs to respect Sylow's theorem. They can mess with finite groups all they want, they can dwarf human intellect every way, but they cannot build a group with 6,300,000,000,000,000,000,000,000 elements that have no subgroup of order 9, they cannot square the circle with ruler and compass, and so forth. What we need to do is to bind them to ethical principles *the same way*. In fact, this is the only truly novel element of our proposal, as the overall goal of somehow endowing AGIs with morality is not new (for a modern summary, see Wallach and Allen (2009)), and as a fundamental ethical precept, the PGC is strongly related to the categorical imperative, which has already received considerable attention as a possible basis of machine morality; see in particular Allen et al. (2000) and Powers (2006).

9.3 ETHICAL RATIONALISM

Our goal is to obtain guarantees of friendliness in a purely deductive fashion. We emphasize at the outset that this is considerably less than what proponents of machine morality generally set out to do: we are not interested in a consistent and complete system of ethics that will tell us in advance what we ought to do in any given circumstances, we are only interested in guidelines that are strong enough to stave off existential threat. In particular, we do not suppose that AGIs need to work toward the benefit of humankind, or to preserve, let alone enhance, the rich fabric of human values. In fact we do not want to presuppose any value system at all, especially as there is a whole school of philosophical thought, starting with Mackie (1977) that takes values to be nonexistent in the first place. Values emerge from Gewirth's analysis, first as entirely subjective valuations of certain things, with no commitment to what these certain things are, and later as necessarily inclusive of certain *rights* the agent must have if it is to be an agent at all. Just as there can be many proofs of the same theorem, there could be many deductive arguments to the desired effect, but so far there seems to be only one, presented in Gewirth (1978), that appears to meet our principal requirement of not using any premiss that lacks empirical evidence. How good is this argument? According to Regis (1984), Gewirth

> gives every appearance of having developed a watertight case, for its arguments are set out with enormous deductive rigor and a frightening dialectical skill. To read Gewirth is to experience the sense of being caught in an ever-tightening net from which all conceivable avenues of escape have been blocked in advance. This is "philosophy as a coercive activity," and Gewirth comes quite close to the extreme of propounding "arguments so powerful they set up reverberations in the brain: if the person refuses to accept the conclusion, he *dies.*" Nevertheless, Gewirth's arguments are not "flashy." They do not proceed by introducing wildly bizarre examples at crucial points; there is no delight in puzzlement for its own sake (…) or contrary-to-fact conditions imposed on imaginary beings hopefully making moral decisions. Rather, Gewirth proceeds by relentlessly piling reason upon reason for thinking that his conclusions are true, and by answering in advance almost every argument for thinking otherwise.

This is not to say that the community of philosophers is uniformly convinced. There are many critical voices, such as Bond (1980) who says that by Gewirth's argument,

> moral evil is reduced to logical error. (…) Gewirth and others like him would turn wickedness into a kind of intellectual incompetence

or Nielsen (1984), who states plainly that Gewirth's central thesis

> that there is a substantive supreme principle of morality, the denial of which is self-contradictory (...) just has to be wrong, and the task (...) is to locate the place or places where such an argument went wrong.

Critics such as Bond may even have it right: the whole point of the enterprise is to demonstrate that wickedness is indeed a form of intellectual incompetence, for if this much is true, the more competent AGIs will restrain the less competent ones from doing wicked things, just as the more competent humans tend to do with the less competent ones (we return to this point in Section 9.4.3). This is not to say that the social process limiting wickedness is perfect, modern history is full of counterexamples from the Third Reich to Cambodia. Obviously, the lower bound on AGI impact cannot be placed below the impact of an exceptional human individual, be their role viewed as positive (say, the appearance of a significant new advocate of nonviolence, such as Mahatma Gandhi) or as negative, such as the appearance of a new dictator.

Whether critics such as Nielsen have it right is another matter entirely. Given the sheer size of Gewirth's argument, 380 pages fully elaborated, with the skeletal version provided in Beyleveld (1992 Part I) running to 60 pages, and given the sophistication of the methods it uses, it demands serious investment of time and energy to fully grasp it, a problem that is faced by ultimately wrong and ultimately right proof attempts alike. The point is not to silence those like Nielsen who are strongly disinclined to accept the argument, it is just as important to seek holes and counterexamples as to strengthen the argument and patch up the holes; the point is to replace philosophical argumentation by formal proof. Here we can only take the first steps in analyzing the argument from the perspective of bounding AGIs. Again we begin with a small example.

9.3.1 Formalizing Philosophical Arguments

Modern artificial theorem proving techniques have largely fulfilled the Leibnizian dream of a *calculus ratiocinator* that would enable symbolic, not just numeric, reasoning by machine. To formalize a philosophical argument, we need just four things: (1) some language describing the expressions we are interested in, (2) some rules for deriving conclusions from premises, (3) some methods to see whether a given rule is applicable, and (4) some methods to see whether premises are met.

There are difficulties at every point: (1) Philosophical arguments are generally given in natural language, as opposed to the formal languages used in logic. As it happens, human-generated mathematical proofs are also published in natural language, and it is well known that a major part of the verification effort lies in translating this language to the language of the theorem prover. (In fact, formalizing Wiles' Fermat proof, comparable in terms of printed pages to the Feit–Thompson proof, has not been accomplished yet; see in particular Hasselink et al. 2006.) If this is already a problem for the highly constrained "natural" language used by mathematicians, it is bound to be even more of a problem for the less constrained language of philosophical discourse. In a similar vein, (2) if the already highly formal deductive style of mathematics is hard to coerce into the mechanical style employed by the theorem prover, the informal deductions employed in philosophy cannot be any easier. As for (3), artificial theorem provers need significant human guidance to find the points where a deductive pattern can be fruitfully matched against the set of true statements already generated from the premises, and the "soft" pattern matching we see in philosophy may pose even more serious problems. (4) On top of this, there is a lack of agreed-upon model theory, and the grounding of philosophical arguments can be surprisingly weak.

Yet in spite of all this, a good argument can be highly compelling. Let us consider the following statement from St. Thomas Aquinas: *even God can't create a mountain without creating a valley.* For Aquinas, this illustrates a stronger statement that omnipotence is limited to the possible, but we need not be actually interested in the notion of omnipotence to appreciate the argument. Let us see how the difficulties enumerated above play out in this case. We take the argument to mean $\forall x \, \text{create}(x, \text{mountain}) \Rightarrow \text{create}(x, \text{valley})$. We do not need to play with the tricky connective *even*, and we do not need a strong notion of *God*. There may be natural language issues, but they do not appear insurmountable. Competent speakers of English (and Latin) will agree that the formulation preserves the hard part: If the formal theorem can be seen to be false, the reasoning behind the natural language statement was weak, and if it can be seen to be true, $\forall x$ must cover even God, so we achieved the effect Aquinas aimed at.

Clearly, a proof cannot be based on the strength of the logical connectives that appear in it. We need some substantive statements about the nature of mountains. We take this as *mountain* $\stackrel{d}{=}$ "land higher than surrounding land." By substitution, if x creates land y higher than the surrounding land, (some) land z lower than y is created by side effect, which is recognized by the soft pattern matching as the *valley*, QED. If the definitions are reasonable, as they are in this case, the conclusion is inevitable. Weak grounding is not a problem, in fact we even gained scope by it, because the same abstract logic applies to electric potential and everywhere else where comparing heights makes sense.

Readers energized by St. Thomas' argument may wish to pursue the ramifications for other kinds of nouns defined by comparative adjectives, for relational nouns such as *parent*, or for plain subsumption (one clearly cannot create a white horse without creating a horse), and so forth. As with any good proof, we soon begin to see that it may have a lot broader scope than what was needed to complete the job at hand.

9.3.2 Outline of the Argument

Gewirth presents his argument *dialectically*, in the original sense of Socratic dialog, rather than in the Hegelian sense of dialectic. This has the advantage that the person the dialectic is aimed at is very soon forced into admitting being a PPA who can *act* with purpose and *reason* rationally. Crucially, the PPA is not assumed to subscribe to any elementary moral prescript, or even the everyday notion of good and bad, let alone good and evil. Such notions, with remarkably specific definitions that make it clear that Gewirth is not just "playing with words," emerge in the course of the argumentation. Following Beyleveld's summary, the main steps of the argument (numbering as in the original) are as follows:

1. I (intend to) do X voluntarily for some purpose E.

2. E is good (by my definition of "good").

3. My freedom and well-being (F&WB) are generically necessary conditions of my agency.

4. My F&WB are necessary goods.

5. I have (maybe nobody else does) a claim right to my F&WB.

6. Other PPAs have a claim right to their F&WB.

7. Every PPA has a claim right to their F&WB.

Gewirth is particularly careful in defending his conclusion against the adeontic viewpoint that there are no claim rights (*ought* statements), the amoralistic viewpoint that I may have claim rights but nobody else does, the consequentialist (classic utilitarian) viewpoint, and so forth. (This is clearly not the place to summarize the debate surrounding the issue, but readers strongly committed to a Moorean notion of a "naturalistic fallacy," or to the "error theory" of Mackie (1977), will find Gewirth (1984) and Stilley (2010) good entry points.) Gewirth is reaching in a deductive fashion some conclusions that have been arrived at in the AGI context by both Omohundro (2008) and Bostrom (2012) by appeal to considerations of fitness: in particular, we see his notion of F&WB as a subset of Omohundro's basic AI drives and Bostrom's instrumental goals. Although cast in a very different (less contemporary but perhaps more rigorous) language, in (3) Gewirth in fact argues for a stronger case than what was made by Omohundro and Bostrom, as he sees F&WB as *generically necessary* conditions of action.

It is, however, not entirely clear that the capacity to reason, in the sense taken for granted by Gewirth, is strictly speaking necessary for AGIs. One may make a strong argument that such capacity will increase fitness, and certainly humans who already have this capacity, even if in a somewhat error-prone fashion as discussed in Section 9.2.3, are unlikely to be seriously threatened by any "intelligence" incapable of abstract reasoning. Notice that reasoning in the abstract, for example, Rybka's chess-playing capability, implies no particular commitment to the kind of symbol manipulation that was central to GOFAI; it simply means that we can use some internal model to make useful predictions about the consequences of various actions.

Bostrom's orthogonality thesis that any level of intelligence can in principle be combined with any final goal is largely borne out by self-inspection. As the current best instantiation of general intelligence, we, humans, are free to choose our final goals. In a more strict sense of orthogonality, intelligence and goals are unlikely to be entirely uncorrelated. In humans we find their goals, as expressed, for example, by choice of career, to be quite predictive of their level of general intelligence, and the negative correlation between criminality and IQ is rather well known, not just at the individual, but also at the state aggregate level (McDaniel 2006). If it can be shown error free, Gewirth's argument will actually trump the orthogonality thesis for the class of AGIs that do have reasoning capabilities sufficiently evolved to comprehend it—we return to this matter in Section 9.4.3.

Given the scrutiny Gewirth's argument already received in the philosophical literature (see, in particular, Regis 1984; Beyleveld 1992), if there are holes in applying the argument to AGIs, they are less likely to come after the premise (1), aimed really at rational human beings, who will be hard put to deny that they have at least *some* intentions to do *something* voluntarily. But a loose coalition of AGIs may even deny the existence of a unified *I* that is the subject of the dialectic (a matter we shall return to in Section 9.4.2), and a superintelligent being may have very good reasons to deny some of the commonsensical assumptions about space and time, actions and consequences, goals and purposes that Gewirth is relying on.

To act with purpose is to act in a voluntary and intentional manner, so a PPA will have at least some notion of some later time. This is already a lot. First, the world must be such that PPAs can have relatively stable dispositions, especially if they can commit to actions that will have to be performed after some delay. An intention to read the next issue of the *Atlantic Monthly* cover to cover implies not only that there will be a next issue (which is not quite a given) but also that I will remember this commitment when the time comes to fulfill it. Second, it entails that we have a means of dealing with failed intentions, because in reality there are such things. Third, we have to be able to stabilize an intention in the sense that X-τ-intends Y at time t (i.e., X intends Y to hold at $t + \tau$) will not be considered true if X is free to change its mind between t and $t + \tau$. It is not that such problems are insurmountable, in fact several solutions are known to the largely analogous Yale shooting problem, but to formalize the entire argument we will need to extend standard action logic (Thielscher 1998; Magnusson 2007) to mental acts and dispositions as well.

Besides a strong reliance on abstract entities such as *purpose, freedom*, or *right*, which can be problematic for a strictly reist model theory, the AGI researcher will immediately note several other characteristics of Gewirth's argument that make formalization a hard task. First, all the reasoning takes place in an ideally resource-unlimited manner: in particular, performing actions and having intentions are largely treated as activities that require no (or negligible) material resources and no (or negligible) time. In reality, many moral conflicts stem from the fact that we need to act before we can think through all the relevant consequences of our actions. This is especially true of deliberate action which may have untold consequences on a large timescale, such as an invention. The inventor of freon could not have possibly foreseen all the consequences. Yet he decided to release the substance to the world, based on imperfect information and a very finite amount of time devoted to reasoning. That this kind of idealization can pose problems was already clear in antiquity "Before he could put into practice something he had heard, the only thing Tzu-lu feared was that he should be told something further" (Analects V.14). The contemporary computer scientist is constitutionally incapable of thinking in a resource-unlimited manner, so the original proof is in a sense better suited to purely mathematical inquiry, with resource bounds added in only afterward.

Another issue, long familiar to students of logic but not particularly touched upon by Gewirth, is the reflexive strength of the deductive system. It is clear that in a world with more than one PPA, there are advantages accruing to each PPA from building internal models of how some other PPA (PPAO) may behave. In particular, if we are smarter than PPAO, we may anticipate its moves and gain all kinds of advantages from doing so. (If we are a lot smarter, we may be able to build a full model and *emulate* PPAO, a matter we shall return to in Section 9.4.3.) We also need to be able to reason about our own reasoning, if only to figure out how PPAO will reason about us. We do not necessarily need fully reflexive reasoning (agents who can reason about reasoning about reasoning about ... their own reasoning), but in a resource-unlimited setting, there seem to be some advantages that an n-fold reflexive PPA will have over an k-fold reflexive PPA for $n > k$. Finally, it should be added that it is not just the epistemic and the deontic modalities that play a significant role in formalizing the argument, but alethic modality is also essential, in that Gewirth aims at strict (categorical, exceptionless, necessary) conclusions at every stage. As we already emphasized at the outset, controlling the power of the modal logic used in formalizing the argument is very much part of the task (see also Section 9.4.3).

But when all is said and done, we do not see any of these difficulties as fatal to the project of formally verifying the argument from (1) to (13). The task is obviously hard and challenging, but the difficulties are not vastly different from those that are faced anyway by those in the AGI community who deal with planning and reasoning. If anything, a shared task like this can bring renewed focus to these efforts.

Because much of contemporary reasoning concerning machine ethics (for a summary, see Muehlhauser and Helm (2012)) is centered on the notions of *utility* and *value*, the considerable simplification brought to the subject by the method of bounds is perhaps worth discussing. First, utility is entirely irrelevant: the argument is fully binding irrespective of the utility function of the agent, if indeed it has one. Second, at this stage we are not at all interested in human values and value systems in general.

What the PGC gives us are rights to F&WB. There may be some slight semantic playing around the edges of really what "freedom" means or "well-being" entails, but the right response is to see which of the possible meanings is actually carried by the formal argument, rather than trying to find the one true meaning, if indeed there is one. This has the somewhat strange and uncomfortable consequence that certain human values will not be carried by the argument, but this is as it should be, given the lack of detailed agreement on what constitutes human value (Yampolskiy 2012). Instead of a mathematically precise and rigorous calculus of moral oughts and ought nots, we end up with a simple statement of *primum nil nocere*. This may be insufficient for fully regulating AGI behavior, but in the final analysis it is about as much as we can reasonably expect from autonomous beings.

9.4 DIFFICULTIES

In this section, we assume that Gewirth's argumentation is not just sound, but entirely flawless, that any sound reasoning agent that grants that it can perform goal-directed action on its own volition will see that the PGC necessarily follows from this very fact. But even if we succeed in the formal verification research program that we sketched in broad strokes, the idea of using ethical rationalism to bound AGI impact still faces some difficulties. From the detached viewpoint that the long time range forces upon us, the problem of *recognizing* PPAs is not just the dialectical problem of AGIs admitting that they are indeed PPAs, the recognition of humans is also problematic—we take up this issue in Section 9.4.1.

Another issue, clearly articulated in Nozick (1981), is that the philosopher can only offer rational reasons to be rational. This is true of a formal verification to an even larger degree, because the philosopher may have some rhetorical resources to move us that the proof checker lacks. But what if an AGI, or a collection of AGIs, refuses to be rational? If the only control on their behavior is some theoretical construct saying that they must respect the rights of others, couldn't they just indulge in all kinds of bad behavior? We turn to this matter in Section 9.4.2.

Finally, a proof presupposes not just the premises, but also the reliability of the logical apparatus it employs. We already alluded to the fact that our discussion is deductive rather than algorithmic, a distinction without a difference as long as we have some form of Curry–Howard correspondence. But philosophical arguments of greater depth so far have only been framed in natural language, where the very existence of a correspondence is unclear. In Section 9.4.3, we take the first, admittedly speculative, steps toward resolving the issue.

9.4.1 Recognizing Humans as PPAs

By Gewirth's argument, we must respect the basic F&WB of other PPAs. He divides *freedom* in subcategories such as "occurrent freedom," the ability of the PPA to control his own particular behaviors by his unforced choice, and "dispositional freedom," his long-range ability to exercise such control. It is precisely because the loss of dispositional freedom (e.g., by imprisonment or enslavement) makes all or most purposive action impossible that Gewirth considers such freedom a generic feature (precondition) of agency.

It is clear that many humans, and not just the prison population, live under conditions so desperate that they cannot realize their potential to purposive agency, yet we must consider them *PPAs*, falling under the scope of the protections offered by PGC. But what about hominids? Modern primate research leaves little doubt that bonobos, chimpanzees, and even orangutans engage in purposive action such as making tools for later use. Our behavior toward animals is strongly contingent on how similar the animal is to us: few people have qualms about poisoning termites or using earthworms as fishing bait. With household pets, our standards are much higher, and in fact cruelty to higher animals is considered both criminal and pathological. A key enabler of our capability to recognize the other as PPAO, mirror neurons (Iacoboni et al. 1999) are hardwired not just in primates but already in birds. We are, it is fair to say, not at all interested in AGIs that are goodwilled but incapable of recognizing us as PPAs.

It is not evident how AGIs lacking in such hardware could recognize humans as PPAs, just as it is unclear that we humans could, or even should, recognize lower life forms from social insects to fish and fowl as (prospective) purposive agents. As long as we see goldfish as having only three seconds of memory (a popular myth now actually debunked), they are just protein-based automata and their F&WB need not be valued. Historically, the easiest way to deny the rights of your opponents is to declare them sub-human—what is to stop some AGI from declaring humanity sub-PPA? Here there are three lines of action, each to be pursued independent of the others.

First, there is broad social critique, so that humanity can get its act together. Although we shall not pursue the issue at any depth here, it should be made clear that animal rights are the least of it: we can begin by considering the kinds of recurrent famines we see in Africa all the time. What makes the situation particularly damning is not that the famine is man-made (the drought is outside human control, but the lack of adequate provisioning is not, cf. Gen. 41:35), but that the very conditions that hamper the delivery of aid are also man-made. Why any higher intelligence should look favorably on a species behaving so badly to its own members is rather unclear.

Second, we may attempt to endow AGIs with PPA detection capabilities. As is clear from Section 9.2.1, this cannot be done by the kind of friend-or-foe devices that are in common use today, for such devices could be easily detached or blindsided. If we follow this route, whatever detection capabilities there are must be both deeply integrated into, and highly valued by, AGIs. Without attempting to speculate further on this matter, we note that in primates the first condition seems to be met directly, as about 13% of the monkey ventral premotor cortex appears to have mirror functions (Kohler et al. 2002), and the second indirectly, as few humans would be willing to give up a significant portion of their brain.

Third, we may attempt to deduct the PPA recognition capability from first principles just like the PGC. Perhaps a lower bound would be sufficient, "if it looks like a PPA and acts like a PPA I assume it's a PPA just to be on the safe side," but for now it is not quite clear on what basis one could attempt a proof that such a discriminative algorithm is not just feasible, but in fact necessary, for a PPA. A possible line of attack may be to demonstrate that a PPA ought, upon reflection, equip oneself with this capability. One thing is for certain: those PPAs that are powerful enough to solve the recognition problem for us by demanding their rights cannot be denied.

9.4.2 Self-Deception

In Section 9.2, we have largely skirted the issue of one or many AGIs, yet it is clear that the bounds placed on an individual will not automatically apply to a larger collective. To the extent there *is* a collective of autonomous but communicating PPAs, we can trust the more intelligent and more powerful members of this collective to restrain the less intelligent ones from doing evil, even if those are still more powerful than humans. Whether the more intelligent (and thus more strongly bound to ethical rationalism) should also be the more powerful is a matter we defer to Section 9.4.3, but we believe that the primary threat is not from fully autonomous agents but rather from semiautonomous ones.

Gewirth's argument creates a bright line between PPAs on the one hand and automata (we use this term here in the sense of "mechanism lacking the essential features of agency," not in the sense of automata theory) on the other: the argument applies only to PPA. Free will is a *sine qua non* of agency: something that performs the exact same steps but without a voluntarily selected goal is not an *agent* but an *instrument*. The distinction may be very hard to make based solely on observing the behavior of an agent, but is very clear proprioceptively: as humans, we consider ourselves having free will. Whether we really do, amplifying quantum indeterminacy to macroscopic action, as suggested by Penrose (1989), or whether we take a compatibilist position, is quite irrelevant here: any machine that fulfills the standard technical definition of nondeterministic computation (Floyd 1967) has the essential features for agency in Gewirth's sense.

Reflection is a *sine qua non* of higher reasoning capability. Therefore, we are less worried about agents that have these capabilities, in that they have the means both to understand, or even discover for themselves, the PGC, and to override other compulsions that would push them in the direction of evil (we use this term indiscriminately for all behaviors that contradict the PGC). The case when the compulsion is too strong for the agent to override falls under a clear moral calculus: such agents are not really agents but instruments and the responsibility lies entirely with their creator.

It is evident that an individual PPA cannot escape responsibility by creating some instrument that will do the dirty work for them. The case of a collective is not so clear-cut. For example, primitive societies that depend on the death penalty will either designate executioners for whom normal moral precepts are assumed to be inoperative or make recourse to stonings, firing squads, execution teams, and other similar tricks to distribute guilt if not causally at least epistemologically. Yet it is clear that anybody who contributes to a causal chain of PGC violation, knowingly or unknowingly, is tainted by this. Society can lift itself to a less primitive level only by the individuals that comprise it taking responsibility. At this point, we run up against the same lower bound that we already discussed in Section 9.3.1—releasing AGIs in the world is no less risky than raising another human. If all else is equal, a body that has some means for dealing with malignancy has a longer life expectancy than one that does not, and a society with the ability to eliminate tainted individuals may also be more resilient. However, this argument only demonstrates that it is prudent to block the morally deficient from acting in society, and says nothing about the means for doing so.

To complete the metaphor, it is not the "killer cells" of an AGI society that we have to worry about, because their own conscience will bound these to the PGC, but something far less science fictional, something we can already observe quite well among humans, self-deception. Situations where "I don't know what took me over" and "I lost control" are part of our everyday experience. We are not just fully rational beings, we are also playing host to many strong internal drives, some inborn, some acquired, and "I know I shouldn't, but" is something that we confront, or suppress, at every slice of cheesecake.

Moral philosophers as diverse as Kant, Kierkegaard, and Sartre have all viewed personal integrity as the capstone that holds the entire moral edifice together. To some extent, this can be explained by the Nozickian desire for a truly compelling argument, for if "the other person is willing to bear the label *irrational* (…) he can skip away happily maintaining his previous belief" (Nozick 1981, p. 4). Kant's *Theory and Practice* dissects the idea

> (…) that a person who lives too much in the world of theory may negligently think that the world in which he actually lives admits of clear application of theory when in fact it does not. Such a person may even come to a distorted view of the world by seeing the world only through the spectacles of his theory—thinking his theory is consistent with the facts because he does not realize that he is unable to accept as a fact anything that is inconsistent with his theory. (Murphy 1998)

In Section 9.2.4, we already discussed that mathematical truth, construed narrowly to exclude long chains of reasoning that can only be performed by machine, is entirely immune to this kind of self-deception in the sense that its failure would demonstrate conclusively that humanity is simply incapable of any kind of reasoning which is coherent with the facts. Although this is not entirely inconceivable (surely this can be one of the six things the White Queen believes before breakfast), the odds are far longer than the 1 in 10^{64} that we took as our baseline.

To the extent self-deception poses a problem for our plan, it is an individual's staying in self-contradicted state, rather than some contradiction between fact and theory, that we need to worry about. Kierkegaard pins his entire theory of the individual on being conscious of the individual's essential responsibility and integrity. To live like an individual, one must have unity. "For he who is not himself a unity is never really anything wholly and decisively; he only exists in an external sense—as long as he lives as a numeral within the crowd, a fraction within the earthly conglomeration.

Alas, how indeed should such a one decide to busy himself with the thought: truthfully to will only one thing!" (Purity of Heart, ch. 13.)

It is remarkable that what we described in the introduction as the relatively mild pain of contradiction is viewed by both Kierkegaard, a deeply Christian thinker, and Sartre, a deeply atheist one, as the greatest blow one can suffer, not willing to be oneself, the condition of *despair*. The human mind is composed of a multitude of somewhat autonomous processes (drives), and one simply cannot let these proceed unchecked, unrecognized, and even overtly denied, if one is to be a moral person or, as these thinkers put it, a person at all. But even if the consequences are as large as existentialism would have it, self-deception is quite frequent and poses a real danger. It is very unlikely that we can construct AGIs that will never be conflicted. We are capable of designing systems that are not crashed by inconsistent data (Belnap 1977), but little effort has gone into systems that can run, in parallel, processes whose goals are inconsistent, or worse yet, run processes whose very existence is denied in the process table. There is a lot to be done both about understanding self-deception in humans (see, in particular, Fingarette (2000)) and in artificial reasoning systems. It may not be necessary to combine this work with the program of verifying rational ethics, for understanding self-deception is a mountain we must climb anyway, but it may prove fruitful to combine the two issues, especially in regard to a critique of tribalism, which we see simply as prolonged societal self-deception that makes it impossible for new members of the society to grow up as rational beings.

9.4.3 The Fitness of Deductive Systems

Understanding Gewirth's argument, if only to the point of being capable of properly challenging it, is already a sign of sophisticated reasoning capabilities. We can easily imagine that highly intelligent purposive agents, such as Attila the Hun, would have had trouble with argumentation at this level of complexity, in fact it is quite unclear how anybody but those familiar with modern Western philosophy could grasp the entire chain of reasoning. Section 9.4.2 left us with the hard question of what is there to stop a higher level AGI from employing lower level "Scourge of God" agents to perform tasks that are incompatible with the PGC. Here we explore a possible solution in terms of yet higher level AGIs. Our remarks, although intended as constructive, must remain rather speculative at this point.

We distinguish three relationships between agents: we say x can *convince* y (about some matter z) or xCy for short, if y will not only acknowledge x's position (about z) as being right but makes it its own in terms of guiding its future voluntary actions. We say that x can *control* y (in regard to z), in short xDy, if x can guarantee that y will act in regard to z in a certain manner even if y voluntarily would not have necessarily done so. Finally, we say that x can *emulate* y, in short xEy (in some respect z, again suppressed in the notation), if x can predict, with absolute certainty, what y would do. Here in "absolute certainty," we include emulation of probabilistic behavior, the case of x using inherently probabilistic devices, if y would do so.

The universally quantified (in z) versions of these three relations are transitive, and all three imply the left-hand side being in some sense stronger than the right-hand side. If y puts overriding value on rationality, xCy implies xDy. If x can clearly anticipate anything y could be doing, x can find the set of arguments that would convince y, so if y can be convinced at all, x is capable of convincing it, meaning xEy also implies xDy. We should add here that it is not just y's propensity to put overriding value on rationality that makes it possible for x to dominate y, if y has a propensity to value empirical evidence, this puts x in the same position as long as x is capable of manipulating the evidence.

We do not have true AGI as of yet, but to the extent we have specialized AI agents, fixing z as it were, humanity clearly has the advantage over these in practical terms. Consider this for the case of computer chess, where AI systems are now several hundred Élö points ahead of the best human players, so humans superficially have no means of winning. But a human player whose only goal is to win against a computer program at all costs can do all kinds of things. He can manipulate the input–output and simply mislead the program into believing that it is playing against a given series of moves although in fact it is playing against some other moves. He can manipulate the low-level addition and multiplication routines that the chess program is relying on. He can directly manipulate the mind-state of the computer, for example, by incrementing some counter in the middle of a search and thus fooling the program into believing that it already considered some alternative. Such steps are obviously unethical, but the situation we are now investigating is precisely the one where the desire to win overrides the ethical imperative.

Classically, theories of logic that meet some basic requirements such as consistency are primarily compared on their *strength*, defined by the variety of elementary classes they can provide first-order axiomatization for. The theories of logic we are interested in must be compared along several dimensions, and strength in the classical sense is not necessarily a primary indicator of the particular notion of strength we are interested in. We will say that a deductive system X is *ahead of* Y in some matter Z if X can prove more from Z than Y. For example, if Y is some calculus of intuitionistic deduction, while X is obtained from Y by the addition of Peirce's formula $\left[(p \rightarrow q) \rightarrow p\right] \rightarrow p$, X will be ahead of Y on some axiom systems Z, and will never be behind it.

The question is not whether a deductive system that is ahead of another is more convincing, for if the deductive apparatus contains objectionable elements, the results obtained by it will also be objectionable. The real issue is whether an AGI that relies on X in the strong sense of accepting X-sanctioned deductions as true even if they are not Y-sanctioned will have any kind of evolutionary advantage over an AGI that relies on Y but not on X. Now, it is not just a formalization of Gewirth's argument in some deductive system Y that we seek, but rather a theorem to the effect that no system X can be ahead of Y unless it also proves the argument. This assures that AGIs respecting the PGC will have an evolutionary advantage over those that do not. If we have such a "son of Lindström" theorem, it provides the enforcement mechanism that secures our main bound even in the face of AGIs that would want to exempt themselves from rational argumentation: more fit AGIs that do respect the PGC.

What is critical is the $Z = \emptyset$ case, the core deductive apparatus, because Gewirth's goal is to derive the PGC without relying on any further axioms. Because Gewirth actually uses modal argumentation at every turn, whether we need something like Barcan's formula in formally reconstructing his reasoning is a key issue. Fortunately, the modal logic used is not deontic but alethic, because the goal is to derive normative statements that have the force of absolute logical necessity. There are many similar bits and pieces of deductive machinery that we will need. Aquinas' argument already relies on the substitutability of equals (Gries and Schneider 1998), and we have emphasized throughout the chapter that the overall power of these pieces needs to be very carefully controlled indeed if we are to have any hope of deriving a "son of Lindström" theorem. Without such a theorem, replicating Gewirth's argument in a formal setting amounts to a study of the design of those AGIs that will voluntarily submit themselves to ethical reasoning, a goal that already makes good sense. With such a theorem we would have even more, because in light of such a theorem the basic AI drives will already make AGIs seek out the high reasoning/high ethics quadrant of Bostrom's orthogonal coordinate space.

It is likely that Attila the Hun cannot be swayed by Gewirth's argument, but as long as there are more powerful intelligences around, they will restrain him because they themselves subscribe to the PGC. Let us suppose that Attila is indeed the Scourge of some higher AGI that could deflect such restraining efforts. But such a higher AGI, God-like as it may appear to us, will respect the PGC (in which case its behavior in letting Attila operate lacks integrity as discussed in Section 9.4.2), or if it does not, AGIs that do can be ahead of it. It should be emphasized in this regard that the PGC is nonnegotiable: there simply cannot be higher reasons, be they prudential, or in the name of some different ethical principle, that are sufficient to deny it. It is precisely this nonnegotiability that a formal proof guarantees: there may be higher intelligences that know a lot more about group theory than I do, in fact there are plenty such people already, but the Sylow theorems bind them just as strongly as they bind me.

A truly general AGI will be much harder to fool than a specialized chess player, because it will be smart enough not to trust external multiplication routines and the like. If it suspects being run in emulation mode, it can cryptographically checksum its state counters—this will not stop external poking but will at least extract some work in return, possibly enough to slow the emulation to a crawl. But as long as xEy is feasible without significant speed loss, clearly x is ahead of y. Evolutionary considerations thus dictate that AGIs always seek out the fastest possible hardware, so as to emulate the old one and use the remaining capacity to improve it. The same considerations dictate that they jealously guard the integrity of their inputs and outputs, and that as long as they strive toward agency, they will also work toward circumventing others' attempts at controlling them. Should they also make themselves immune to reasoning? Remarkably, here the opposite strategy makes more sense, for as long as xCy makes x more fit, it is in the best interest of y to adopt the reasoning offered by x.

As is clear from the foregoing, any AGI expecting to reach a high level of fitness will find it prudent to expend some effort toward tamper-proofing its environment, its perceptual and motor systems, and its internal logic. Once these efforts are deemed successful (and they can never be completely successful in the material universe in that arbitrarily large gamma-ray bursts can always reset some part of memory), we can equate an AGI with its deductive system. It is therefore a reasonable long-term goal to attempt to compare and evolve AGIs in a proof checker environment, but it is clear that the short-term proof-checking goal outlined in Section 9.3 is already very ambitious. A key issue is that systems of deduction are not at all first-class entities—rather, they get hardwired in the proof checker.

9.5 CONCLUSIONS

In the history of ideas, ethical precepts are traditionally attributed to the sages. Variants of the PGC go back to Confucius "Do not impose on others what you yourself do not desire" (Analects XII 2), Buddha "Hurt not others in ways that you yourself would find hurtful" (Udana-Varga 5.18), Jesus "So in everything, do to others what you would have them do to you, for this sums up the Law and the Prophets" (Matt. 7:12), and Muhammad "No one of you shall become a true believer until he desires for his brother what he desires for himself" (Sahih Al-Bukhari), and can be found in almost any sacred book from the *Mahabharata* "Do not do to others what would cause pain if done to you" (5.1517) to the *Shayest Na-Shayest* "Not to do unto others all that which is not well for one's self" (13.29). This tradition assumes that ethics is divinely inspired, and thus ethical laws carry a special, transcendent authority.

Another view, characteristic of the Enlightenment, and given modern form in Rawls (1971), takes morals to be the result of a social contract. Closely related is the historical view, which takes them to be the result of a long societal process that begins with "folk law" (Renteln and Dundes 1995). Modern research extends this to prehistory based on the observation that not just humans but primates already come with inherited moral traits such as compassion (de Waal 1997, 2009), and in Section 9.4.1, we already pointed at one issue, recognition of PPAs, that seems to rely on some form of hardware support. To the extent collaborative behavior can be advantageous even in a purely goal-directed setting (Munoz de Côte et al. 2010; Waser 2012), in due time we can expect the PGC to emerge directly under evolutionary pressure. As Bayles (1968) notes:

> It would seem that [egoism] would often result in severe competition between people, since each person would be out to get the most good for himself, and this might involve his depriving others. However, serious defenders of egoism, e.g. Hobbes and Spinoza, have generally held that upon a rational examination of the human situation it appears one best promotes his own interest by co-operating with others.

One thing that seems to stand in the way of an evolutionary justification of morals is the variety of instinctive behaviors we see in animals. Because many of these are strikingly egoistic at both the individual and the species level, it seems the evolutionary pressure toward collaboration is considerably less than that for improved sensory and motor systems. Also, it seems that evolution bequeaths to more complex organisms a whole set of drives that are often in conflict. The pioneers of cybernetics were greatly worried that rats will, under certain experimental conditions (starved both for sex and for food), prefer sex to exploration, exploration to food, and food to sex (McCulloch 1945). Although such circular preferences in public opinion were already known to Condorcet, the fact that an organism as simple as a rat (today we have more respect for the internal complexity of rodents than was common in the postwar period) can already harbor contradictory drives was seen at the time as fatal to any attempt at modeling the obviously more complex human behavior (let alone the presumably even more complex AGI behavior) by any utility function.

Ethical rationalism offers a way out of the conundrum of highly evolved but immoral behavioral patterns such as brood parasitism, in that it relies on agency and reflective reasoning, facilities that are largely absent from animals other than hominids and perhaps cetaceans. As we emphasized at the outset, the essence of the method of bounds is to trade in precision for reliability. Evolution will necessarily proceed in a haphazard, probabilistic fashion, but the argument Gewirth deploys steers clear of any form of relying on probabilistic or deterministic, computable/hypercomputable or uncomputable, utility function. Also, it is worth emphasizing that the bound will apply to singletons as well, even if they are not subject to ordinary evolutionary pressures.

Recently, Goertzel and Pitt (2012) have laid out a plan to endow AGIs with morality by means of carefully controlled machine learning. Much as we are in agreement with their goals, we remain skeptical about their plan meeting the plain safety engineering criteria laid out at the beginning. Instead, we suggest that the essence of AGIs is their reasoning facilities, and it is the very logic of their being that will compel them to behave in a moral fashion. Therefore, we see theorem provers as the natural habitat of AGIs until we are satisfied they can be let loose. The real nightmare scenario (called "all bets are off" in Bukatin (2000)) is one where there is no "son of Lindström" theorem, but some humans find it advantageous to strongly couple themselves to AGIs, with no guarantees against self-deception. Modern society is constructed so that the selectional pressure toward higher intelligence is immense, witness the spread of smart drugs, so the Faustian bargain of (surgically?) coupling oneself to a mind-expanding

AGI may prove irresistible. On this centenary, we feel that chartering a Turing Police of the kind described by Gibson in 1984, another pregnant date, may not be too far off.

ACKNOWLEDGMENTS

I thank Michael Bukatin (Brandeis University, Waltham, MA), Abram Demski (University of Southern California, Los Angeles, CA), William Hibbard (SSEC, University of Wisconsin–Madison, Madison, WI), and Luke Muehlhauser (Singularity Institute) for cogent criticism. This work was supported by OTKA grant #82333.

APPENDIX: THE SIZE OF THE EXISTENTIAL THREAT

Our understanding of the dangers facing humankind is rather limited. We only have a few, imperfectly understood data points, and estimates of the death toll of even such recent and well-documented events as the Cambodian genocide, or the ongoing Iraq conflict, are not accurate within 10%. Nevertheless, we can single out some points in the geological record where mass extinctions indubitably took place. A good example is the Ordovician–Silurian extinction event that occurred some 443.7 million years ago: all main phyla were decimated and nearly half of the genera (49% according to Rohde and Muller 2005) became extinct. The causes of this and similar extinctions are ill-understood, with continental drift, meteorite impact, and gamma-ray bursts standing out as the most widely accepted hypotheses. Needless to say, understanding the causes of this magnitude is in no way tantamount to controlling them, in spite of the widespread belief, sustained by movies such as *Armageddon*, that there is nothing that a few heroic people and a few good nukes will not take care of.

When designing radioactive equipment, a reasonable guideline is to limit emissions to several orders of magnitude below the natural background radiation level, so that human-caused dangers are lost in the noise compared to the preexisting threat we must live with anyway. Here we take the "big five" extinction events that occurred within the past half billion years as background. Assuming a mean time of 10^8 years between mass extinctions and 10^9 victims in the next one yields an annualized death rate of 10, comparing quite favorably to the reported global death rate of ~500 for contact with hornets, wasps, and bees (ICD-9-CM E905.3), not to speak of death from famine, wars, and preventable diseases, which have several orders of magnitude higher death tolls (though the annualized rates are declining, see Pinker (2011)). Martel (1997) estimates a considerably higher annualized death rate of 3500 from meteorite impacts alone (she does not consider continental drift or gamma-ray bursts), but the internal logic of safety engineering demands we seek a lower bound, one that we must put up with no matter what strides we make in redistribution of food, global peace, or health care.

Let us define *existential threat* as some AGI (individual or collective) pushing the wrong button. Current computers operate in the gigahertz range, so they can perform roughly 10^9 operations per second, or about 10^{17} operations annually. Clock speeds will no doubt continue to increase, and there is no easily defensible upper bound in sight. Therefore, we use the Planck limit and assume at most 10^{56} logic operations per year per processor. For an AGI with a finger on the button to be *less* of an existential threat than the threat from the astronomical background by some safety factor $m = 10^s$, it needs a guaranteed failure rate of no more than one in 10^{56+s} logic operations. If there is not one AGI but several, we can use the computational capacity of the universe, estimated by Lloyd (2002) as 10^{120} operations. These numbers compare rather starkly with the best that humanity can currently manage, the Long Now Foundation's clocks with a planned lifetime of 3×10^{11} seconds.

BIBLIOGRAPHY

Allen, C., Varner, G., and Zinser, J. (2000), "Prolegomena to any future artificial moral agent," *Journal of Experimental & Theoretical Artificial Intelligence*, 12(3), 251–261.

Aoyama, T., Hayakawa, M., Kinoshita, T., and Nio, M. (2007), "Revised value of the eighth-order contribution to the electron g-2," *Physical Review Letters*, 99(11), 110406.

Appel, K., and Haken, W. (1976), "A proof of the four color theorem," *Discrete Math*, 16(2), 179–180.

Aschbacher, M. (2004), "The status of the classification of the finite simple groups," *Notices of the AMS*, 51(7), 736–740.

Aschbacher, M., and Smith, S. (2004a), *The Classification of Quasithin Groups*, Vol. I, American Mathematical Society, Providence, RI.

Aschbacher, M., and Smith, S. (2004b), *The Classification of Quasithin Groups*, Vol. II, American Mathematical Society, Providence, RI.

Bayles, M.D. (1968), *Contemporary Utilitarianism*, Vol. 644, Anchor Books.

Belnap, N.D. (1977), "How a computer should think," in *Contemporary Aspects of Philosophy*, ed. G. Ryle, Oriel Press, Newcastle upon Tyne, pp. 30–56.

Bender, H., and Glauberman, G. (1995), *Local Analysis for the Odd Order Theorem*, Vol. 188, Cambridge University Press.

Beyleveld, D. (1992), *The Dialectical Necessity of Morality: An Analysis and Defense of Alan Gewirth's Argument to the Principle of Generic Consistency*, University of Chicago Press.

Bond, E.J. (1980), "Reply to Gewirth," *Metaphilosophy*, 11(1), 70–75.

Bostrom, N. (2012), "The superintelligent will: Motivation and instrumental rationality in advanced artificial agents," *Minds and Machines*, 22, 71–75.

Bukatin, M. (2000), *Singularity Is More Radical Than We Think*, Vol. http://www. cs.brandeis.edu/~bukatin/singularity.html, Accessed November 28, 2012.

Dancoff, S.M. (1939), "On radiative corrections for electron scattering," *Physical Review*, 55(10), 959.

De Waal, F. (1997), *Good Natured: The Origins of Right and Wrong in Humans and Other Animals*, Harvard University Press.

De Waal, F. (2009), *Primates and Philosophers: How Morality Evolved*, Princeton University Press.

Feit, W., and Thompson, J.G. (1963), "Solvability of groups of odd order," *Pacific Journal of Mathematics*, 13(3), 775–1029.

Feynman, R. (1985), *QED. The Strange Theory of Matter and Light*, Princeton, NJ: Princeton University Press.

Fingarette, H. (2000), *Self-Deception*, University of California Press.

Floyd, R.W. (1967), "Nondeterministic algorithms," *Journal of the ACM*, 14(4), 636–644.

Gewirth, A. (1978), *Reason and Morality*, University of Chicago Press.

Gewirth, A. (1984), "Replies to my critics," in *Gewirth's Ethical Rationalism*, ed. E. Regis Jr., University of Chicago Press, pp. 192–256.

Gibson, W. (1984), *Neuromancer*, Ace Science Fiction.

Goertzel, B., and Pitt, J. (2012), "Nine ways to bias open-source AGI toward friendliness," *Journal of Evolution and Technology*, 22, 116–131.

Gonthier, G. (2008), "Formal proof-the-four-color theorem," *Notices of the AMS*, 55(11), 1382–1393.

Gorenstein, D. (1982), *Finite Simple Groups: An Introduction to Their Classification*, Plenum Press, New York.

Gries, D., and Schneider, F.B. (1998), *Formalizations of Substitution of Equals for Equals*, Technical Report, CS Department, Cornell University.

Iacoboni, M., Woods, R.P., Brass, M., Bekkering, H., Mazziotta, J.C., and Rizzolatti, G. (1999), "Cortical mechanisms of human imitation," *Science*, 286(5449), 2526–2528.

Kempe, A.B. (1879), "On the geographical problem of the four colours," *American Journal of Mathematics*, 2(3), 193–200.

Kinoshita, T., and Nio, M. (2006), "Improved $\alpha 4$ term of the electron anomalous magnetic moment," *Physical Review D*, 73(1), 013003.

Knies, R. (2012), *Theorem Proof Gains Acclaim*, http://research.microsoft.com/en-us/news/features/gonthierproof-101112.aspx.

Kohler, E., Keysers, C., Umilta, M.A., Fogassi, L., Gallese, V., and Rizzolatti, G. (2002), "Hearing sounds, understanding actions: Action representation in mirror neurons," *Science*, 297(5582), 846–848.

Lánczos, C. (1964), "A precision approximation of the gamma function," *Journal of the Society for Industrial & Applied Mathematics, Series B: Numerical Analysis*, 1(1), 86–96.

Lloyd, S. (2002), "Computational capacity of the universe," *Physical Review Letters*, 88(23), 237901.

Mackie, J.L. (1977), *Ethics: Inventing Right and Wrong*, Penguin.

Magnusson, M. (2007), *Deductive Planning and Composite Actions in Temporal Action Logic*, Linköping University.

Martel, L. M.V. (1997), "Damage by impact. The case at Meteor Crater, Arizona," *Planetary Science Research Discoveries*.

McCulloch, W.S. (1945), "A heterarchy of values determined by the topology of nervous nets," *Bulletin of Mathematical Biophysics*, 7, 89–93.

McDaniel, M.A. (2006), "Estimating state IQ: Measurement challenges and preliminary correlates," *Intelligence*, 34, 607–619.

Mohr, P.J., Taylor, B.N., and Newell, D.B. (2012), "CODATA recommended values of the fundamental physical constants: 2010."

Muehlhauser, L. (2012), *AI Risk Bibliography 2012*, The Singularity Institute.

Muehlhauser, L., and Helm, L. (2012), "The singularity and machine ethics," in *The Singularity Hypotheses: A Scientific and Philosophical Assessment*, eds. A. Eden, J. Sraker, J.H. Moor, and E. Steinhart, Springer.

Munoz de Côte, E., Chapman, A., Sykulski, A., and Jennings, N. (2010), "Automated planning in adversarial repeated games," in *Proceedings of the 26th Conference on Uncertainty in Artificial Intelligence*, pp. 376–383.

Murphy, J.G. (1998), "Kant on theory and practice," in *Character, Liberty and Law: Kantian Essays in Theory and Practice*, Kluwer.

Nederpelt, R.P., Geuvers, J.H., and de Vrijer, R.C., eds. (1994), *Selected Papers on Automath*, North Holland.

Nielsen, K. (1984), "Against ethical rationalism," in *Gewirth's Ethical Rationalism: Critical Essays with a Reply by Alan Gewirth*, University of Chicago Press, pp. 59–83.

Nozick, R. (1981), *Philosophical Explanations*, Harvard University Press.

Omohundro, S. (2008), "The basic AI drives," in *Proceedings of the 1st Artificial General Intelligence Conference*, eds. P. Wang, B. Goertzel, and S. Franklin, IOS Press.

Ord, T. (2002), *Hypercomputation: Computing More than the Turing Machine*, http://arxiv.org/ftp/math/papers/0209/0209332.pdf.

Penrose, R. (1989), *The Emperor's New Mind*: Concerning computers, minds, and the laws of physics, Oxford University Press.

Perelman, G. (1994), "Manifolds of positive Ricci curvature with almost maximal volume," *Journal of the American Mathematical Society*, 7, 299–305.

Peterfalvi, T. (2000), *Character Theory for the Odd Order Theorem*, Vol. 272, Cambridge University Press.

Pinker, S. (2011), *The Better Angels of Our Nature: Why Violence Has Declined*, Viking.

Powers, T.M. (2006), "Prospects for a Kantian machine," *Intelligent Systems, IEEE*, 21(4), 46–51.

Rawls, J. (1971), *A Theory of Justice*, Harvard University Press.

Regis, E. (1984), *Gewirth's Ethical Rationalism: Critical Essays with a Reply by Alan Gewirth*, University of Chicago Press.

Renteln, A., and Dundes, A. (1995), *Folk Law: Essays in the Theory and Practice of Lex Non Scripta*, Vol. 2.

Rohde, R.A., and Muller, R.A. (2005), "Cycles in fossil diversity," *Nature*, 434(7030), 208–210.

Stilley, S. (2010), *Natural Law Theory and the "Is"-"Ought" Problem: A Critique of Four Solutions*, http://epublications.marquette.edu/dissertations_mu/57.

Taylor, B.N., Parker, W.W., and Langenberg, D.N. (1969), "Determination of e/h, using macroscopic quantum phase coherence in superconductors: Implications for quantum electro-dynamics and the fundamental physical constants," *Reviews of Modern Physics*, 41(3), 375.

Thielscher, M. (1998), "Reasoning about actions: Steady versus stabilizing state constraints," *Artificial Intelligence*, 104(1), 339–355.

Tomonaga, S. (1966), "Development of quantum electrodynamics," *Physics Today*, 19, 25.

Wallach, W., and Allen, C. (2009), *Moral Machines: Teaching Robots Right from Wrong*, Oxford University Press.

Waser, M. (2012), *Backward Induction: Rationality or Inappropriate Reductionism? Part 1*, http://transhumanity.net/articles/entry/backward-induction-rationality-or-inappropriate-reductionism-part-1.

Wiles, A. (1995), "Modular elliptic curves and Fermat's last theorem," *Annals of Mathematics*, 443–551.

Yampolskiy, R.V. (2012), "Artificial intelligence safety engineering: Why machine ethics is a wrong approach," in *Philosophy and Theory of Artificial Intelligence*, *SAPERE 5*, ed. V. Müller, Springer, pp. 389–396.

Yampolskiy, R.V., and Fox, J. (2012), "Safety engineering for artificial general intelligence," *Topoi*, 1–10.

Yudkowsky, E. (2001), *Creating Friendly AI 1.0: The Analysis and Design of Benevolent Goal Architectures*, Technical Report, The Singularity Institute, San Francisco, CA.

CHAPTER 10

Ethics of Brain Emulations

Anders Sandberg

CONTENTS

ABSTRACT Whole brain emulation attempts to achieve software intelligence by copying the function of biological nervous systems into software. This chapter aims at giving an overview of the ethical issues of the brain emulation approach, and analyze how they should affect responsible policy for developing the field. Animal emulations have uncertain moral status, and a principle of analogy is proposed for judging treatment of virtual animals. Various considerations of developing and using human brain emulations are discussed.

10.1 INTRODUCTION

Whole brain emulation (WBE) is an approach to achieve software intelligence by copying the functional structure of biological nervous systems into software. Rather than attempting to understand the high-level processes underlying perception, action, emotions, and intelligence, the approach assumes that they would emerge from a sufficiently close imitation of the low-level neural functions, even if this is done through a software process (Merkle, 1989; Sandberg & Bostrom, 2008).

Although the philosophy (Chalmers, 2010), impact (Hanson, 2008), and feasibility (Sandberg, 2013) of brain emulations have been discussed, little analysis of the ethics of the project so far has been done. The main questions of this chapter are to what extent brain emulations are moral patients and what new ethical concerns are introduced as a result of brain emulation technology.

10.1.1 Brain Emulation

The basic idea is to take a particular brain, scan its structure in detail at some resolution, and construct a software model of the physiology that is so faithful to the original that, when run on appropriate hardware, it will have an internal causal structure that is essentially the same as the original brain. All relevant functions on some level of description are present and higher level functions supervene from these.

Although at present being an unfeasibly ambitious challenge, the necessary computing power and various scanning methods are rapidly developing. Large-scale computational brain models are a very active research area, at present reaching the size of mammalian nervous systems (Markram, 2006; Djurfeldt et al., 2008; Eliasmith et al., 2012; Preissl et al., 2012). WBE can be viewed as the logical endpoint of current trends in computational neuroscience and systems biology. Obviously the eventual feasibility depends on a number of philosophical issues (physicalism, functionalism, non-organicism) and empirical facts (computability, scale separation, detectability, scanning, and simulation tractability) that cannot be predicted beforehand; WBE can be viewed as a program trying to test them empirically (Sandberg, 2013).

Early projects are likely to merge data from multiple brains and studies, attempting to show that this can produce a sufficiently rich model to produce nontrivial behavior but not attempting to emulate any particular individual. However, it is not clear that this can be carried on indefinitely: higher mammalian brains are organized and simultaneously individualized through experience, and linking parts of different brains is unlikely to produce functional behavior. This means that the focus is likely to move to developing a "pipeline" from brain to executable model, where ideally an individually learned behavior of the original animal is demonstrated by the resulting emulation.

Although WBE focuses on the brain, a realistic project will likely have to include a fairly complex body model in order to allow the emulated nervous system to interact with a simulated or real world, as well as the physiological feedback loops that influence neural activity.

At present the only known methods able to generate complete data at cellular and subcellular resolution are destructive, making the scanned brain nonviable. For a number of reasons, it is unlikely that nondestructive methods will be developed any time soon (Sandberg & Bostrom, 2008, Appendix E).

In the following, I will assume that WBE is doable, or at least does not suffer enough roadblocks to preclude attempting it, in order to examine the ethics of pursuing the project.

10.2 VIRTUAL LAB ANIMALS

The aim of brain emulation is to create systems that closely imitate real biological organisms in terms of behavior and internal causal structure. Although the ultimate ambitions may be grand, there are many practical uses of intermediate realistic organism simulations. In particular, emulations of animals could be used instead of real animals for experiments in education, science, medicine, or engineering. Opponents of animal testing often argue that much of it is excessive and could be replaced with simulations. Although the current situation is debatable, in a future where brain emulations are possible it would seem that this would be true: by definition, emulations would produce the same kind of results as real animals.

However, there are three problems:

- Brain emulations might require significant use of test animals to develop the technology.

- Detecting that something is a perfect emulation might be impossible.

- An emulation might hold the same moral weight as a real animal by being sentient or a being with inherent value.

10.2.1 Need for Experiments

Developing brain emulation is going to require the use of animals. It would be necessary not only for direct scanning into emulations, but also in various experiments for gathering the necessary understanding of neuroscience, testing scanning modalities, and comparing the real and simulated animals. In order to achieve a useful simulation, we need to understand at least one relevant level of the real system well enough to recreate it, otherwise the simulation will not produce correct data.

What kind of lab animals would be suitable for research in brain emulation and how would they be used? At present neuroscientists use nearly all model species, from nematode worms to primates. Typically there are few restrictions on research on invertebrates (with the exception of cephalopods). Although early attempts are likely to aim at simple, well-defined nervous systems such as the nematode *Caenorhabditis elegans*, *Lymnaea stagnalis* (British pond snail), or *Drosophila melanogaster* (fruit fly), much of the neuroscience and tool development will likely involve standard vertebrate lab animals such as mice, either for *in vitro* experiments with tissue pieces or for *in vivo* experiments attempting to map neural function to properties that can be detected. The nervous system of invertebrates also differ in many ways from the mammalian nervous system; although they might make good test benches for small emulations, it is likely that the research will tend to move toward small mammals, hoping that successes can be scaled up to larger brains and bodies. The final stages in animal brain emulation before moving on to human emulation would likely involve primates, raising the strongest animal protection issues. In theory this stage might be avoidable if the scaling up from smaller animal brains toward humans seems smooth enough, but this would put a greater risk on the human test subjects.

Most "slice and dice" scanning (where the brain is removed, fixated, and then analyzed) avoids normal animal experimentation concerns because there is no experiment done on the living animal itself, just tissue extraction. This is essentially terminal anesthesia ("unclassified" in UK classification of suffering). The only issue here is the prescanning treatment, whether there is any harm to the animal in its life coming to an end, and whether *in silico* suffering possible.

However, developing brain emulation techniques will likely also involve experiments on living animals, including testing whether an *in vivo* preparation behaves like an *in vitro* and an *in silico* model. This will necessitate using behaving animals in ways that could cause suffering. The amount of such research needed is at present hard to estimate. If the non-organicism assumption of WBE is correct, most data gathering and analysis will deal with low-level systems such as neuron physiology and connectivity rather than the whole organism; if all levels are needed, then the fundamental feasibility of WBE is cast into question (Sandberg, 2013).

10.2.2 What Can We Learn from Emulations?

The second problem is equivalent to the current issue of how well animal models map onto human conditions, or more generally how much models and simulations in science reflect anything about reality.

The aim is achieving structural validity (Zeigler, 1985; Zeigler, Praehofer, & Kim, 2000) that the emulation reflects how the real system operates. Unfortunately this might be impossible to prove: there could exist hidden properties that only very rarely come into play that are not represented. Even defining meaningful and observable measures of success is nontrivial when dealing with higher order systems (Sandberg, 2013). Developing methods and criteria for validating neuroscience models is one of the key requirements for WBE.

One of the peculiar things about the brain emulation program is that unlike many scientific projects the aim is not directly full understanding of the system that is being simulated. Rather, the simulation is used as a verification of our low-level understanding of neural systems and is intended as a useful tool. Once successful, emulations become a very powerful tool for further investigations (or valuable in themselves). Before that stage, the emulation does not contribute much knowledge about the full system. This might be seen as an argument against undertaking the WBE project: the cost and animals used are not outweighed by returns in the form of useful scientific knowledge. However, sometimes very risky projects are worth doing because they promise very large eventual returns (consider the Panama Canal) or might have unexpected but significant spin-offs (consider the Human Genome Project). Where the balance lies depends on both how the evaluations are made and the degree of long-term ambition.

10.2.3 What Is the Moral Status of an Emulation?

The question what moral consideration we should give to animals lies at the core of the debate about animal experimentation ethics. We can pose a similar question about what moral claims emulations have on us. Can they be wronged? Can they suffer?

Indirect theories argue that animals do not merit moral consideration, but the effect of human actions on them does matter. The classic example is Kantian theories, where animals lack moral autonomy and hence are not beings whose interests morally count. Our duties toward them are merely indirect duties toward humanity. Being cruel to animals harms our own humanity:

> Our duties towards animals are merely indirect duties towards humanity. Animal nature has analogies to human nature, and by doing our duties to animals in respect of manifestations of human nature, we indirectly do our duty to humanity …. We can judge the heart of a man by his treatment of animals.

> REGAN & SINGER, 1989, PP. 23–24

By this kind of indirect account, the nature of the emulation does not matter: if it is cruel to pinch the tail of biological mice, the same cruel impulse is present in pinching the simulated tail of an emulated mouse. It is like damaging an effigy: it is the intention behind doing damage that is morally bad, not the damage. Conversely, treating emulations well might be like treating dolls well: it might not be morally obligatory but its compassionate.

A different take on animal moral considerability comes from social contract or feminist ethics, arguing against the individualist bias they perceive in the other theories. What matters is not intrinsic properties but the social relations we have with animals.

> Moral considerability is not an intrinsic property of any creature, nor is it supervenient on only its intrinsic properties, such as its capacities. It depends, deeply, on the kind of relations they can have with us

> ANDERSON, 2004, P. 289

If we have the same kind of relations to an emulated animal as a biological animal, they should presumably be treated similarly. Because successful emulations (by assumption) also have the same capacity to form reciprocal relations, this seems likely.

Another large set of theories argues that the interests of animals do count morally due to intrinsic properties. Typically they are based on the sentience of animals giving them moral status: experiences of pleasure or suffering are morally relevant states no matter what system experiences them. Whether animals are sentient or not is usually estimated from the argument from analogy, which supports claims of consciousness by looking at similarities between animals and human beings. Species membership is not a relevant factor. These theories differ on whether human interests still can trump animal interests or whether animals actually have the same moral status as human beings. For the present purpose, the important question is whether software emulations can have sentience, consciousness, or the other properties these theories ground moral status on.

Animal rights can be argued on other grounds than sentience, such as animals having beliefs, desires, and self-consciousness of their own, and hence having inherent value and rights as subjects of a life that has inherent value (Regan, 1983). Successfully emulated animals would presumably behave in similar ways: the virtual mouse will avoid virtual pain; the isolated social animal will behave in a lonely fashion. Whether the mere behavior of loneliness or pain avoidance is an indication of a real moral interest even when we doubt it is associated with any inner experience is problematic: most accounts of moral patienthood take experience as fundamental, because that actually ties the state of affairs to a value, the welfare of something. But theories of value that ascribe value to nonagents can of course allow nonconscious software as a moral patient (e.g., having value by virtue of its unique complexity).

To my knowledge, nobody has yet voiced concern that existing computational neuroscience simulations could have aversive experiences. In fact, the assumption that simulations do not have phenomenal consciousness is often used to motivate such research:

> Secondly, one of the more obvious features of mathematical modelling is that it is not invasive, and hence could be of great advantage in the study of chronic pain. There are major ethical problems with the experimental study of chronic pain in humans and animals. It is possible to use mathematical modelling to test some of the neurochemical and neurophysiological features of chronic pain

without the use of methods which would be ethically prohibitive in the laboratory or clinic. Stembach has observed "Before inflicting pain on humans, can mathematical or statistical modelling provide answers to the questions being considered?" (p. 262) (53). We claim that mathematical modelling has the potential to add something unique to the armamentarium of the pain researcher.

BRITTON & SKEVINGTON, 1996, P. 1139

To some degree, this view is natural because typical computational simulations contain just a handful of neurons. It is unlikely that so small systems could suffer.[*] However, the largest simulations have reached millions or even billions of neurons: we are reaching the numbers found in brains of small vertebrates that people do find morally relevant. The lack of meaningful internal structure in the network probably prevents any experience from occurring, but this is merely a conjecture.

Whether machines can be built to have consciousness or phenomenological states has been debated for a long time, often as a version of the strong artificial intelligence (AI) hypothesis. At one extreme, it has been suggested that even thermostats have simple conscious states (Chalmers, 1996), making phenomenal states independent of higher level functions, whereas opponents of strong AI have commonly denied the possibility of any machine (or at least software) mental states. See Gamez (2008) for a review of some current directions in machine consciousness.

It is worth noting that there are cognitive scientists who produce computational models they consider able to have consciousness (as per their own theories).[†]

---

[*] However, note the Small Network Argument (Herzog et al., 2007): "... for each model of consciousness there exists a minimal model, i.e., a small neural network, that fulfills the respective criteria, but to which one would not like to assign consciousness." Mere size of the model is not a solution: there is little reason to think that 10^{11} randomly connected neurons are conscious, and appeals to the right kind of complexity of interconnectedness run into the argument again. One way out is to argue that fine-grained consciousness requires at least mid-sized systems: small networks only have rudimentary conscious contents (Taylor, 2007). Another one is to bite the bullet and accept, if not panpsychism, that consciousness might exist in exceedingly simple systems.

Assigning even a small probability to the possibility of suffering or moral importance to simple systems leads to far bigger consequences than just making neuroscience simulations suspect. The total number of insects in the world is so great that if they matter morally even to a tiny degree, their interests would likely overshadow humanity's interests. This is by no means a *reductio ad absurdum* of the idea: it could be that we are very seriously wrong about what truly matters in the world.

[†] See, for example, the contributions in the theme issue of *Neural Networks*, Volume 20, Issue 9, November 2007.

Consider the case of Rodney Cotterill's CyberChild, a simulated infant controlled by a biologically inspired neural network and with a simulated body (Cotterill, 2003). Within the network, different neuron populations corresponding to brain areas such as the cerebellum, brainstem nuclei, motor cortices, sensory cortex, hippocampus, and amygdala are connected according to an idealized mammalian brain architecture with learning, attention, and efference copy signals. The body model has some simulated muscles and states such as levels of blood glucose, milk in the stomach, and urine in the bladder. If the glucose level drops too much, it "expires." The simulated voice and motions allow it to interact with a user, trying to survive by getting enough milk. Leaving aside the extremely small neural network (20 neurons per area), it is an ambitious project. This simulation does attempt to implement a model of consciousness, and the originator was hopeful that there was no fundamental reason why consciousness could not ultimately develop in it.

However, were the CyberChild conscious, it would have a very impoverished existence. It would exist in a world of mainly visual perception, except for visceral inputs, "pain" from full nappies, and hunger. Its only means of communication is crying and the only possible response is the appearance (or not) of a bottle that has to be maneuvered to the mouth. Even if the perceptions did not have any aversive content, there would be no prospect of growth or change.

This is eerily similar to Thomas Metzinger's warning:

> What would you say if someone came along and said, "Hey, we want to genetically engineer mentally retarded human infants! For reasons of scientific progress we need infants with certain cognitive and emotional deficits in order to study their postnatal psychological development—we urgently need some funding for this important and innovative kind of research!" You would certainly think this was not only an absurd and appalling but also a dangerous idea. It would hopefully not pass any ethics committee in the democratic world. However, what today's ethics committees *don't* see is how the first machines satisfying a minimally sufficient set of constraints for conscious experience could be just *like* such mentally retarded infants. They would suffer from all kinds of functional and representational deficits too. But they would now also subjectively experience those deficits. In addition, they would have no political lobby—no representatives in *any* ethics committee.

METZINGER, 2003, P. 621

He goes on arguing that we should ban all attempts to create or even risk the creation artificial systems that have phenomenological self-models. Although views on what the particular criterion for being able to suffer is might differ between different thinkers, it is clear that the potential for suffering software should be a normative concern. However, as discussed in mainstream animal rights ethics, other interests (such as human interests) can sometimes be strong enough to allow animal suffering. Presumably such interests (if these accounts of ethics are correct) would also allow for creating suffering software.

David Gamez (2005) suggests a probability scale for machine phenomenology, based on the intuition that machines built along the same lines as human beings are more likely to have conscious states than other kinds of machines. This scale aims to quantify how likely a machine is to be *ascribed* to be able to exhibit such states (and to some extent, address Metzinger's ethical concerns without stifling research). In the case of WBE, the strong isomorphism with animal brains gives a fairly high score.[*] Arrabales, Ledezma, and Sanchis (2010), however, suggests a scale for the estimation of the potential degree of consciousness based on architectural and behavioral features of an agent; again a successful or even partial WBE implementation of an animal would by definition score highly (with a score dependent on species). The actual validity and utility of such scales can be debated, but insofar they formalize intuitions about the argument from analogy about potential mental content they show that WBE at least has significant *apparent* potential of being a system that has states that might make it a moral patient. WBE is different from entirely artificial software in that it deliberately tries to be as similar as possible to morally considerable biological systems, and this should make us more ethically cautious than with other software.

Much to the point of this section, Dennett (1978) has argued that creating a machine able to feel pain is nontrivial, to a large extent in the incoherencies in our ordinary concept of pain. However, he is not against the possibility in principle:

> If and when a good physiological sub-personal theory of pain is developed, a robot could in principle be constructed to instantiate

---

[*] For an electrophysiological WBE model, the factors are FW1, FM1, FN4, and AD3, with rate, size, and time slicing possibly ranging over the whole range. This produces a weighting ranging between 10^{-5} and 0.01, giving an ordinal ranking 170-39 out of 812. The highest weighting beats the neural controlled animat of DeMarse et al., a system containing real biological neurons controlling a robot.

it. Such advances in science would probably bring in their train wide-scale changes in what we found intuitive about pain, so that the charge that our robot only suffered what we artificially called pain would lose its persuasiveness. In the meantime (if there were a cultural lag) thoughtful people would refrain from kicking such a robot.

DENNETT, 1978, P. 449

From the eliminative materialist perspective, we should hence be cautious about ascribing or not ascribing suffering to software, because we do not (yet) have a good understanding of what suffering is (or rather, what the actual underlying component that is morally relevant is). In particular, successful WBE might indeed represent a physiological subpersonal theory of pain, but it might be as opaque to outside observers as real physiological pain.

The fact that at present there does not seem to be any idea of how to solve the hard problem of consciousness or how to detect phenomenal states seems to push us in the direction of suspending judgment:

> Second, there are the arguments of Moor (1988) and Prinz (2003), who suggest that it may be indeterminable whether a machine is conscious or not. This could force us to acknowledge the possibility of consciousness in a machine, even if we cannot tell for certain whether this is the case by solving the hard problem of consciousness.

GAMEZ, 2008, P. 892

Although the problem of animal experience and status is contentious, the problem of emulated experience and status will by definition be even more contentious. Intuitions are likely to strongly diverge and there might not be any empirical observations that could settle the differences.

10.2.4 Principle of Assuming the Most

What to do in a situation of moral uncertainty about the status of emulations?* It seems that a safe strategy would be to make the most cautious assumption:

---

* Strictly speaking, we are in a situation of moral uncertainty about what ethical system we ought to follow in general and factual uncertainty about the experiential status of emulations. But being sure about one and not the other one still leads to a problematic moral choice. Given the divergent views of experts on both questions, we should also not be overly confident about our ability to be certain in these matters.

> *Principle of assuming the most (PAM)*: Assume that any emulated system could have the same mental properties as the original system and treat it correspondingly.

The mice should be treated the same in the real laboratory as in the virtual. It is better to treat a simulacrum as the real thing than to mistreat a sentient being. This has the advantage that many of the ethical principles, regulations, and guidance in animal testing can be carried over directly to the pursuit of brain emulation.

This has some similarity to the principle of substrate nondiscrimination ("If two beings have the same functionality and the same conscious experience, and differ only in the substrate of their implementation, then they have the same moral status.") (Bostrom & Yudkowsky, 2014) but does not assume that the conscious experience is identical. However, if one were to reject the principle of substrate nondiscrimination on some grounds, then it seems that one could also reject PAM because one does have a clear theory of what systems have moral status. However, this seems to be a presumptuous move given the uncertainty of the question.

Note that once the principle is applied, it makes sense to investigate in what ways the assumptions can be sharpened. If there are reasons to think that certain mental properties are *not* present, they overrule the principle in that case. An emulated mouse that does not respond to sensory stimuli is clearly different from a normal mouse. It is also relevant to compare to the right system. For example, the CyberChild, despite its suggestive appearance, is not an emulation of a human infant but at most an etiolated subset of neurons in a generic mammalian nervous system.

It might be argued that this principle is too extreme, and that it forecloses much of the useful pain research discussed by Britton and Skevington (1996). However, it is agnostic on whether there exist overruling human interests. That is left for the ethical theory of the user to determine, for example, using cost–benefit methods. Also, as discussed in the following, it might be quite possible to investigate pain systems without phenomenal consciousness.

10.2.5 Ameliorating Virtual Suffering

PAM implies that unless there is evidence to the contrary, we should treat emulated animals with the same care as the original animal. This means in most cases that practices that would be impermissible in the physical lab are impermissible in the virtual lab.

Conversely, counterparts to practices that reduce suffering such as analgesic practices should be developed for use on emulated systems. Many of the best practices discussed in the work of Schofield (2002) can be readily implemented: brain emulation technology by definition allows parameters of the emulation that can be changed to produce the same functional effect as the drugs have in the real nervous system. In addition, pain systems can in theory be perfectly controlled in emulations (e.g., by inhibiting their output), producing "perfect painkillers." However, this is all based on the assumption that we understand what is involved in the experience of pain: if there are undiscovered systems of suffering careless, research can produce undetected distress.

It is also possible to run only part of an emulation, for example, leaving out or blocking nociceptors, the spinal or central pain systems, or systems related to consciousness. This could be done more exactly (and reversibly) than in biological animals. Emulations can also be run for very brief spans of time, not allowing any time for a subjective experience. But organisms are organic wholes with densely interacting parts: just like in real animal ethics there will no doubt exist situations where experiments hinge upon the whole behaving organism, including its aversive experiences.

It is likely that early scans, models, and simulations will often be flawed. Flawed scans would be equivalent to animals with local or global brain damage. Flawed models would introduce systemic distortions, ranging from the state of not having any brain to abnormal brain states. Flawed simulations (broken off because of software crashes) would correspond to premature death (possibly repeated, with no memory—see later). Viewed in analogy with animals, it seems that the main worry should be flawed models producing hard-to-detect suffering.

Just like in animal research, it is possible to develop best practices. We can approximate enough of the inner life of animals from empirical observations to make some inferences; the same process is in principle possible with emulations to detect problems peculiar to their state. In fact, the transparency of an emulation to data gathering makes it easier to detect certain things such as activation of pain systems or behavioral withdrawal, and backtrack their source.

10.2.6 Quality of Life

An increasing emphasis is placed not just on lack of suffering among lab animals but on adequate quality of life. What constitutes adequate is itself a research issue. In the case of emulations, the problem is that quality of life presumably requires both an adequate body and an adequate environment for the simulated body to exist in.

The virtual reality world of an emulated nematode or snail is likely going to be very simple and crude even compared to their normal Petri dish or aquarium, but the creatures are unlikely to consider that bad. But as we move up among the mammals, we will get to organisms that have a quality of life. A crude virtual reality world might suffice for testing the methods, but would it be acceptable to keep a mouse, cat, or monkey in an environment that is too bare for any extended time? Worse, can we know in what ways it is too bare? We have no way of estimating the importance rats place on smells, and whether the smell in the virtual cage is rich enough to be adequate. The intricacy of body simulations also matters: how realistic does a fur have to feel to simulated touch to be adequate?

I estimate that the computational demands of running a very realistic environment are possible to meet and not terribly costly compared to the basic simulation (Sandberg & Bostrom, 2008, pp. 76–78). However, modeling the *right* aspects requires a sensitive understanding of the lifeworlds of animals we might simply be unable to reliably meet. However, besides the ethical reasons to pursue this understanding, there is also a practical need: it is unlikely emulations can be properly validated unless they are placed in realistic environments.

10.2.7 Euthanasia

Most regulations of animal testing see suffering as the central issue, and hence euthanasia as a way of reducing it. Some critics of animal experimentation however argue that an animal life holds intrinsic value, and ending it is wrong.

In the emulation case, strange things can happen, because it is possible (due to the multiple realizability of software) to create multiple instances of the same emulation and to terminate them at different times. If the end of the identifiable life of an instance is a wrong, then it might be possible to produce large number of wrongs by repeatedly running and deleting instances of an emulation even if the experiences during the run are neutral or identical.

Would it matter if the emulation was just run for a millisecond of subjective time? During this time, there would not be enough time for any information transmission across the emulated brain, so presumably there could not be any subjective experience. Accounts of value of life built upon being a subject of a life would likely find this unproblematic: the brief emulations do not have a time to be subjects; the only loss might be to the original emulation if this form of future is against its interests.

Conversely, what about running an emulation for a certain time, making a backup copy of its state, and then deleting the running emulation only to have it replaced by the backup? In this case, there would be a break in continuity of the emulation that is only observable on the outside and a loss of experience that would depend on the interval between the backup and the replacement. It seems unclear that anything is lost if the interval is very short. Regan (1983) argues that the harm of death is a function of the opportunities of satisfaction it forecloses; in this case, it seems that it forecloses the opportunities envisioned by the instance while it is running, but it is balanced by whatever satisfaction can be achieved during that time.

Most concepts of the harmfulness of death deal with the irreversible and identity-changing aspects of the cessation of life. Typically, any reversible harm will be lesser than an irreversible harm. Because emulation makes several of the potential harms of death (suffering while dying, stopping experience, bodily destruction, changes of identity, cessation of existence) completely or partially reversible, it actually reduces the sting of death.

In situations where there is a choice between the irreversible death of a biological being and an emulation counterpart, the PAM suggests that we ought to play it safe: they might be morally equivalent. The fact that we might legitimately doubt whether the emulation is a moral patient does not mean it has a value intermediate between the biological being and nothing, but rather that the actual value is *either* full or none, we just do not know which. If the case is the conversion of the biological being into an emulation, we are making a gamble that we are not destroying something of value (under the usual constraints in animal research of overriding interests, or perhaps human autonomy in the case of a human volunteer).

However, the reversibility of many forms of emulation death may make it cheaper. In a lifeboat case (Regan, 1983), should we sacrifice the software? If it can be restored from backup, the real loss will be just the lost memories since last backup and possibly some freedom. Death forecloses fewer opportunities to emulations.

It might of course be argued that the problem is not ending emulations, but the fundamental lack of respect for a being. This is very similar to human dignity arguments, where humans are assumed to have intrinsic dignity that can never be removed, yet it can be gravely disrespected. The emulated mouse might not notice anything wrong, but we know it is treated in a disrespectful way.

There is a generally accepted view that animal life should not be taken wantonly. However, emulations might weaken this: It is easy and painless to end an emulation, and it might be restored with equal ease with no apparent harm done. If more animals are needed, they can be instantiated up to the limits set by available hardware. Could emulations hence lead to a reduction of the value of emulated life? Slippery slope arguments are rarely compelling; the relevant issues rather seem to be that the harm of death has been reduced and that animals have become (economically) cheap. The moral value does not hinge on these factors but on the earlier discussed properties. That does not mean we should ignore risks of motivated cognition changing our moral views, but the problem lies in complacent moral practice rather than emulation.

10.2.8 Conclusion

Developing animal emulations would be a long-running, widely distributed project that would require significant animal use. This is not different from other major neuroscience undertakings. It might help achieve replacement and reduction in the long run, but could introduce a new morally relevant category of sentient software. Due to the uncertainty about this category, I suggest a cautious approach: it should be treated as the corresponding animal system absent countervailing evidence. Although this would impose some restrictions on modeling practice, these are not too onerous, especially given the possibility of better-than-real analgesia. However, questions of how to demonstrate scientific validity, quality of life, and appropriate treatment of emulated animals over their "life span" remain.

10.3 HUMAN EMULATIONS

Brain emulation of humans raises a host of extra ethical issues or sharpens the problems of proper animal experimentation.

10.3.1 Moral Status

The question of moral status is easier to handle in the case of human emulations than in the animal case because they can report back about their state. If a person who is skeptical of brain emulations being conscious or having free will is emulated and, after due introspection and consideration, changes their mind, then that would seem to be some evidence in favor of emulations actually having an inner life. It would actually not prove anything stronger than that the processes where a person changes their mind are correctly emulated and that there would be some disconfirming evidence in the emulation. It could still be a lacking consciousness and be a functional philosophical zombie (assuming this concept is even coherent).

If philosophical zombies existed, it seems likely that they would be regarded persons (at least in the social sense). They would behave like persons, they would vote, they would complain and demand human rights if mistreated, and in most scenarios there would not be any way of distinguishing the zombies from the humans. Hence, if emulations of human brains work well enough to exhibit humanlike behavior rather than mere humanlike neuroscience, legal personhood is likely to eventually follow, despite misgivings of skeptical philosophers.*

10.3.2 Volunteers and Emulation Rights

An obvious question is volunteer selection. Is it possible to give informed consent to brain emulation? The most likely scanning methods are going to be destructive, meaning that they end the biological life of the volunteer or would be applied to postmortem brains.

In the first case, given the uncertainty about the mental state of software, there is no way of guaranteeing that there will anything "after," even if the scanning and emulation are successful (and of course the issues of personal identity and continuity). Hence, volunteering while alive is essentially equivalent to assisted suicide with an unknown probability of "failure." It is unlikely that this will be legal on its own for quite some time even in liberal jurisdictions: suicide is increasingly accepted as a way of escaping pain, but suicide for science is not regarded as an acceptable reason.† Some terminal patients might yet argue that they wish to use this particular form of "suicide" rather than a guaranteed death, and would seem to have autonomy on their side. An analogy can be made

---

* Francis Fukuyama, for example, argues that emulations would lack consciousness or true emotions, and hence lack moral standing. It would hence be morally acceptable to turn them off at will (Fukyama, 2002, pp. 167–170). In light of his larger argument about creeping threats to human dignity, he would presumably see working human WBE as an insidious threat to dignity by reducing us to mere computation. However, exactly what factor to base dignity claims on is if in anything more contested than what to base moral status on; see, for example, Bostrom (2009) for a very different take on the concept.

† The Nuremberg code states: "No experiment should be conducted, where there is an a priori reason to believe that death or disabling injury will occur; except, perhaps, in those experiments where the experimental physicians also serve as subjects." But self-experimentation is unlikely to make high risk studies that would otherwise be unethical ethical. Some experiments may produce so lasting harm that they cannot be justified for any social value of the research (Miller and Rosenstein, 2008).

to the use of experimental therapies by the terminally ill, where concerns about harm must be weighed against uncertainty about the therapy, and where the vulnerability of the patient makes them exploitable (Falit & Gross, 2008).

In the second case, postmortem brain scanning, the legal and ethical situation appears easier. There is no legal or ethical person in existence, just the preferences of a past person and the rules for handling anatomical donations. However, this also means that a successful brain emulation based on a person would exist in a legal limbo. Current views would hold it to be a possession of whatever institution performed the experiment rather than a person.*

Presumably, a sufficiently successful human brain emulation (especially if it followed a series of increasingly plausible animal emulations) would be able to convince society that it was a thinking, feeling being with moral agency and hence entitled to various rights. The PAM would support this: even if one were skeptical of whether the being was "real," the moral risk of not treating a potential moral agent well would be worse than the risk of treating nonmoral agents better than needed. Whether this would be convincing enough to have the order of death nullified and the emulation regarded as the same *legal* person as the donor is another matter, as is issues of property ownership.

The risk of ending up a nonperson and possibly being used against one's will for someone's purposes, ending up in a brain-damaged state, or ending up in an alien future, might not deter volunteers. It certainly does not deter people signing contract for cryonic preservation today, although they are fully aware that they will be stored as nonperson anatomical donations and might be revived in the future with divergent moral and social views. Given that the alternative is certain death, it appears to be a rational choice for many.

---

* People involved in attempts at preserving legally dead but hopefully recoverable patients are trying to promote recognition of some rights of stored individuals. See, for instance, the Bill of Brain Preservation Rights (Brain Preservation Foundation, 2013). Specifically, it argues that persons in storage should be afforded similar rights to living humans in temporally unconscious states. This includes ensuring quality medical treatment and long-term storage, but also revival rights ("The revival wishes of the individual undergoing brain preservation should be respected, when technically feasible. This includes the right to partial revival (memory donation instead of identity or self-awareness revival), and the right to refuse revival under a list of circumstances provided by the individual before preservation.") and legal rights allowing stored persons to retain some monetary or other assets in trust form that could be retrieved upon successful revival. This bill of rights would seem to suggest similar right for stored stored emulations.

10.3.3 Handling of Flawed, Distressed Versions

While this problem is troublesome for experimental animals, it becomes worse for attempted human emulations. The reason is that unless the emulation is so damaged that it cannot be said to be a mind with any rights, the process might produce distressed minds that are rightsholders yet have existences not worth living, or lack the capacity to form or express their wishes. For example, they could exist in analogous to persistent vegetative states, dementia, schizophrenia, aphasia, or have ongoing very aversive experience. Many of these ethical problems are identical to current cases in medical ethics.

One view would be that if we are ethically forbidden from pulling the plug of a counterpart biological human, we are forbidden from doing the same to the emulation. This might lead to a situation where we have a large number of emulation "patients" requiring significant resources, yet not contributing anything to refining the technology nor having any realistic chance of a "cure."

However, brain emulation allows a separation of cessation of experience from permanent death. A running emulation can be stopped and its state stored, for possible future reinstantiation. This leads to a situation where at least the aversive or meaningless experience is stopped (and computational resources freed up), but which poses questions about the rights of the now frozen emulations to eventual revival. What if they were left on a shelf forever, without ever restarting? That would be the same as if they had been deleted. But do they in that case have a right to be run at least occasionally, despite lacking any benefit from the experience?

Obviously methods of detecting distress and agreed on criteria for termination and storage will have to be developed well in advance of human brain emulation, likely based on existing precedents in medicine, law, and ethics.

Persons might write advance directives about the treatment of their emulations. This appears equivalent to normal advance directives, although the reversibility of local termination makes pulling the plug less problematic. It is less clear how to handle wishes to have more subtly deranged instances terminated. Although a person might not wish to have a version of himself or herself with a personality disorder become his or her successor, at the point where the emulation comes into being, it will potentially be a moral subject with a right to its life, and might regard its changed personality as the right one.

10.3.4 Identity

Personal identity is likely going to be a major issue, both because of the transition from an original unproblematic human identity to successor identity/identities that might or might not be the same, and because software minds can potentially have multiple realizability. The discussion about how personal identity relates to successor identities on different substrates is already extensive and will be foregone here. See, for instance, Chalmers (2010).

Instantiating multiple copies of an emulation and running them as separate computations is obviously as feasible as running a single one. If they have different inputs (or simulated neuronal noise), they will over time diverge into different persons, who have not just a shared past but at least initially very similar outlooks and mental states.

Obviously multiple copies of the same original person pose intriguing legal challenges. For example, contract law would need to be updated to handle contracts where one of the parties is copied—does the contract now apply to both? What about marriages? Are all copies descended from a person legally culpable of past deeds occurring before the copying? To what extent does the privileged understanding copies have of each other affect their suitability as witnesses against each other? How should votes be allocated if copying is relatively cheap and persons can do "ballot box stuffing" with copies? Do copies start out with equal shares of the original's property? If so, what about inactive backup copies? And so on. These issues are entertaining to speculate upon and will no doubt lead to major legal, social, and political changes if they become relevant.

From an ethical standpoint, if all instances are moral agents, then the key question is how obligations, rights, and other properties carry over from originals to copies and whether the existence of copies changes some of these. For example, making a promise "I will do X" is typically meant to signify that the future instance of the person making the promise will do X. If there are two instances, it might be enough that one of them does X (whereas promising *not* to do X might morally oblige both instances to abstain). But this assumes that the future instances acknowledge the person doing the promise as their past self—a perhaps reasonable assumption, but one which could be called into question if there is an identity affecting transition to brain emulation in between (or any other radical change in self-identity).

Would it be moral to voluntarily undergo very painful and/or lethal experiments given that at the end that suffering copy would be deleted and replaced with a backup made just after making the (voluntarily and fully informed) decision to participate? It seems that current views on scientific self-experimentation do not allow such behavior on the grounds that there are certain things it is never acceptable to do for science. It might be regarded as a combination of the excessive suffering argument (there are suffering so great that no possible advance in knowledge can outweigh its evil) and a human dignity argument (it would be a practice that degrades the dignity of humans). However, the views on what constitutes unacceptable suffering and risk have changed historically and are not consistent across domains. Performance artists sometimes perform acts that would be clearly disallowed as scientific acts, yet the benefit of their art is entirely subjective (Goodall, 1999). It might be that as the technology becomes available, boundaries will be adjusted to reflect updated estimates of what is excessive or lacks dignity, just as we have done in many other areas (e.g., reproductive medicine, transplant medicine).

10.3.5 Time and Communication

Emulations will presumably have experience and behave on a timescale set by the speed of their software. The speed a simulation is run relative to the outside world can be changed, depending on available hardware and software. Current large-scale neural simulations are commonly run with slowdown factors on the order of a thousand, but there does not seem to be any reason precluding emulations running faster than real-time biological brains.*

Nick Bostrom and Eliezer Yudkowsky (2014) have argued for a principle of subjective rate of time: "In cases where the duration of an experience is of basic normative significance, it is the experience's subjective duration that counts." By this account frozen states do not count at all. Conversely, very fast emulations can rapidly produce a large amount of positive or negative values if they are in extreme states: they might count for more in utilitarian calculations.

Does human emulation have a right to real time? Being run at a far faster or slower rate does not matter as long as an emulation is only interacting with a virtual world and other emulations updating at the same speed. But when interacting with the outside world, speed matters. A divergent

---

* Axons typically have conduction speeds between 1 and 100 m/s, producing delays between a few and a hundred milliseconds in the brain. Neurons fire at less than 100 Hz. Modern CPUs are many orders of magnitude faster (in the gigahertz range) and transmit signals at least 10% of the speed of light. A millionfold speed increase does not seem implausible.

clockspeed would make communication with people troublesome or impossible. Participation in social activities and meaningful relationships depend on interaction and might be made impossible if they speed past faster than the emulation can handle. A very fast emulation would be isolated from the outside world by lightspeed lags and from biological humans by their glacial slowness. It hence seems that insofar emulated persons are to enjoy human rights (which typically hinge on interactions with other persons and institutions), they need to have access to real-time interaction, or at least "disability support" if they cannot run fast enough (e.g., very early emulations with speed limited by available computer power).

By the same token, this may mean that emulated humans have a right to contact with the world outside their simulation. As Nozick's experience machine demonstrates, most people seem to want to interact with the "real world," although that might just mean the shared social reality of meaningful activity rather than the outside physical world. At the very least emulated people would need some "I/O rights" for communication within their community. But because the virtual world is contingent upon the physical world and asymmetrically affected by it, restricting access only to the virtual is not enough if the emulated people are to be equal citizens of their wider society.

10.3.6 Vulnerability

Brain emulations are extremely vulnerable by default: the software and data constituting them and their mental states can be erased or changed by anybody with access to the system on which they are running. Their bodies are not self-contained and their survival is dependent upon hardware they might not have causal control over. They can also be subjected to undetectable violations such as illicit copying. From an emulation perspective, software security is identical to personal security.

Emulations also have a problematic privacy situation, because not only their behavior can be perfectly documented by the very system they are running on, but also their complete brain states are open for inspection. Whether that information can be interpreted in a meaningful way depends on future advances in cognitive neuroscience, but it is not unreasonable to think that by the time human emulations exist, many neural correlates of private mental states will be known. This would put them in a precarious situation.

These considerations suggest that the ethical way of handling brain emulations would be to require strict privacy protection of the emulations and that the emulated persons had legal protection or ownership of the hardware on which they are running, because it is in a sense their physical bodies. Some technological solutions such as encrypted simulation or tamper-resistant special purpose hardware might help. How this can be squared with actual technological praxis (e.g., running emulations as distributed processes on rented computers in the cloud) and economic considerations (what if an emulation ran out of funds to pay for its upkeep?) remains to be seen.

10.3.7 Self-Ownership

Even if emulations are given personhood, they might still find the ownership of parts of themselves to be complicated. It is not obvious that an emulation can claim to own the brain scan that produced it: it was made at a point in time where the person did not legally exist. The process might also produce valuable intellectual property, for example, useful neural networks that can be integrated in nonemulation software to solve problems, in which case the matter of who has a right to the property and proceeds from it, emerge.

This is not just an academic question: ownership is often important for developing technologies. Investors want to have returns on their investment, and innovators want to retain control over their innovations. This was apparently what partially motivated the ruling in *Moore v. Regents of the University of California* that a patient did not have property rights to cells extracted from his body and turned into lucrative products (Gold, 1998). This might produce pressures that work against eventual self-ownership for the brain emulations.

Conversely, essential subsystems of the emulation software or hardware may be licensed or outright owned by other parties. Does right to life trump or self-ownership property ownership? At least in the case of the first emulations, it is unlikely that they would have been able to sign any legal contracts, and they might have a claim. However, the owners might still want fair compensation. Would it be acceptable for owners of computing facilities to slow down or freeze nonpaying emulations? It seems that the exact answer depends on how emulation self-ownership is framed ethically and legally.

10.3.8 Global Issues and Existential Risk

The preliminary work that has been done on the economics and social impact of brain emulation suggests that they could be a massively disruptive force. In particular, simple economic models predict that copy-able human capital produces explosive economic growth and (emulated) population increase but also wages decreasing toward Malthusian levels (Hanson, 1994, 2008). Economies that can harness emulation technology well might have a huge strategic advantage over latecomers.

WBE could introduce numerous destabilizing effects, such as increase of inequality, groups that become marginalized, disruption of existing social power relationships, and the creation of opportunities to estab-lish new kinds of power, the creation of situations in which the scope of human rights and property rights is poorly defined and subject to dispute, and particularly strong triggers for racist and xenophobic prejudices, or vigorous religious objections. Although all these factors are speculative and depend on the details of the world and WBE emergence scenario, they are a cause for concern.

An often underappreciated problem is existential risk: the risk that humanity and all Earth-derived life go extinct (or suffers a global, perma-nent reduction in potential or experience) (Bostrom, 2002). Ethical analy-sis of the issue shows that reducing existential risk tends to take strong precedence over many other considerations (Bostrom, 2003, 2013). Brain emulations have a problematic role in this regard. On the one hand, they might lead to various dystopian scenarios, but on the other hand, they might enable some very good outcomes.

As discussed above, the lead-up to human brain emulation might be very turbulent because of arms races between different groups pursuing this potentially strategic technology; fearful other groups would reach the goal ahead of them. This might continue after the breakthrough, now in the form of wild economic or other competition. Although the technol-ogy itself is not doing much, the sheer scope of what it *could* do leads to potential war.

It could also be that competitive pressures or social drift in a society with brain emulation leads to outcomes where value is lost. For exam-ple, wage competition between copyable minds may drive wages down to Malthusian levels and produce beings only optimized for work, spending all available resources on replication, or gradual improvements in emula-tion efficiency lose axiologically valuable traits (Bostrom, 2004). If emu-lations are zombies, humanity, tempted by cybernetic immortality, may gradually trade away its consciousness. These may be evolutionary attrac-tors that may prove inescapable without central coordination: each step toward the negative outcome is individually rational.

However, there are at least four major ways emulations that might lower the risks of Earth-originating intelligence going extinct:

1. The existence of nonbiological humans would ensure at least partial protection from some threats: there is no biological pandemic that can wipe out software. Of course, it is easy to imagine a digital disaster, for example, an outbreak of computer viruses that wipe out the brain emulations. But that threat would not affect the biological humans. By splitting the human species into two, the joint risks are significantly reduced. Clearly threats to the shared essential infrastructure remain, but the new system is more resilient.

2. Brain emulations are ideally suited for colonizing space and many other environments where biological humans require extensive life support. Avoiding carrying all eggs in one planetary basket is an obvious strategy for strongly reducing existential risk. Besides existing in a substrate-independent manner where they could be run on computers hardened for local conditions, emulations could be transmitted digitally across interplanetary distances. One of the largest obstacles of space colonization is the enormous cost in time, energy, and reaction mass needed for space travel: emulation technology would reduce this.

3. Another set of species risks accrues from the emergence of machine superintelligence. It has been argued that successful AI is potentially extremely dangerous because it would have radical potential for self-improvement yet possibly deeply flawed goals or motivation systems. If intelligence is defined as the ability to achieve one's goals in general environments, then superintelligent systems would be significantly better than humans at achieving their goals—even at the expense of human goals. Intelligence does not strongly prescribe the nature of goals (especially in systems that might have been given top-level goals by imperfect programmers). Brain emulations gets around part of this risk by replacing the *de novo* machine intelligence with a copy of the relatively well-understood human intelligence. Instead of getting potentially very rapidly upgradeable software minds with nonhuman motivation systems, we get messy emulations that have human motivations. This slows the "hard takeoff" into superintelligence and allows existing, well-tested forms

of control over behavior—norms, police, economic incentives, and political institutions—to act on the software. This is by no means a guarantee: emulations might prove to be far more upgradeable than we currently expect; motivations might shift from human norms; speed differences and socioeconomial factors that may create turbulence; and the development of emulations might also create spin-off AI.

4. Emulation allows exploration of another part of the space of possible minds, which might encompass states of very high value (Bostrom, 2008).

Unfortunately, these considerations do not lend themselves to easy comparison. They all depend on somewhat speculative possibilities, and their probabilities and magnitude cannot easily be compared. Rather than giving a rationale for going ahead or for stopping WBE, they give reasons for assuming WBE will—were it to succeed—matter enormously. The value of information helping determining the correct course of action is hence significant.

10.4 DISCUSSION: SPECULATION OR JUST BEING PROACTIVE?

Turning back from these long-range visions, we get to the mundane but essential issues of research ethics and the ethics of ethical discourse.

Ethics matters because we want to do good. In order to do that, we need to have some ideas of what the good is and how to pursue it in the right way. It is also necessary for establishing trust with the rest of society—not just as public relations or a way of avoiding backlashes, but in order to reap the benefit of greater cooperation and useful criticism.

There is a real risk of both overselling and dismissing brain emulation. It has been a mainstay of philosophical thought experiments and science fiction for a long time. The potential impact for humanity (and to currently living individuals hoping for immortality) could be enormous. Unlike *de novo* AI, it appears possible to benchmark progress toward brain emulation, promising a more concrete (if arduous) path toward software intelligence. It is a very concrete research goal that can be visualized, and it will likely have a multitude of useful spin-off technologies and scientific findings no matter its eventual success.

Yet these stimulating factors also make us ignore the very real gaps in our knowledge, the massive difference between the current technology and the technology we can conjecture we need, and the foundational uncertainty about whether the project is even feasible. This lack of knowledge easily leads to a split into a camp of enthusiasts who assume that the eventual answers will prove positive and a camp of skeptics who dismiss the whole endeavor. In both camps, motivated cognition will filter evidence to suit their interpretation, producing biased claims and preventing actual epistemic progress.

There is also a risk that ethicists work hard on inventing problems that are not there. After all, institutional rewards go to ethicists that find high-impact topics to pursue, and hence it makes sense arguing that whatever topic one is studying is of higher impact than commonly perceived. Alfred Nordmann has argued that much debate about human enhancement is based on "if … then…" ethical thought experiments where some radical technology is first assumed, and then the ethical impact explored in this far-fetched scenario. He argued that this wastes limited ethical resources on flights of fantasy rather than the very real ethical problems we have today (Nordmann, 2007).

Nevertheless, considering potential risks and their ethical impacts is an important aspect of research ethics, even when dealing with merely possible future radical technologies. Low-probability, high-impact risks do matter, especially if we can reduce them by taking proactive steps in the present. In many cases, the steps are simply to gather better information and have a few preliminary guidelines ready if the future arrives surprisingly early. Although we have little information in the present, we have great leverage over the future. When the future arrives, we may know far more, but we will have less ability to change it.

In the case of WBE, the main conclusion of this chapter is the need for computational modelers to safeguard against software suffering. At present this would merely consist of being aware of the possibility, monitor the progress of the field, and consider what animal protection practices can be imported into the research models when needed.

ACKNOWLEDGMENTS

I thank Håkan Andersson, Peter Eckersly, Toby Ord, Catarina Lamm, Stuart Armstrong, Vincent C. Müller, Gaverick Matheny, and Randal Koene for many stimulating discussions helping to shape this chapter.

REFERENCES

Anderson, E. (2004). Animal rights and the values of nonhuman life. In Sunstein, C. R., & Nussbaum, M. (Eds.), *Animal Rights: Current Debates and New Directions* (p. 289). Oxford: Oxford University Press.

Arrabales, R., Ledezma, A., & Sanchis, A. (2010). ConsScale: A pragmatic scale for measuring the level of consciousness in artificial agents. *Journal of Consciousness Studies*, 17(3–4), 131–164.

Bostrom, N. (2002). Existential risks: Analyzing human extinction scenarios and related hazards. *Journal of Evolution and Technology*, 9(1), 1–37.

Bostrom, N. (2003). Astronomical waste: The opportunity cost of delayed technological development. *Utilitas*, 15(3), 308–314.

Bostrom, N. (2004). The future of human evolution. In Tandy, C. (Ed.), *Death and Anti-Death: Two Hundred Years after Kant, Fifty Years after Turing* (pp. 339–371). Palo Alto, CA: Ria University Press.

Bostrom, N. (2008). Why I want to be a posthuman when I grow up. In Gordijn, B., & Chadwick, R. (Eds.), *Medical Enhancement and Posthumanity* (pp. 107–137). New York: Springer.

Bostrom, N. (2009). Dignity and enhancement. *Contemporary Readings in Law and Social Justice*, 2, 84–115.

Bostrom, N. (2013). Existential risk prevention as global priority. *Global Policy*, 4(1), 15–31.

Bostrom, N., & Yudkowsky, E. (2014). The ethics of artificial intelligence. In Ramsey, W., & Frankish, K. (Eds.), *The Cambridge Handbook of Artificial Intelligence*. Cambridge: Cambridge University Press, 316–334.

Brain Preservation Foundation. (2013). *Bill of Preservation Rights*. Retrieved from http://www.brainpreservation.org/content/preservation-rights (downloaded on March 1, 2013).

Britton, N. F., & Skevington, S. M. (1996). On the mathematical modelling of pain. *Neurochemical Research*, 21(9), 1133–1140.

Chalmers, D. (1996). *The Conscious Mind: In Search of a Fundamental Theory*. New York: Oxford University Press.

Chalmers, D. (2010). The singularity: A philosophical analysis. *Journal of Consciousness Studies*, 17(9/10), 7–65.

Cotterill, R. (2003). CyberChild: A simulation test-bed for consciousness studies. In Holland, O. (Ed.), *Machine Consciousness*. Exeter: Imprint Academic.

Dennett, D. C. (1978). Why you can't make a computer that feels pain. *Synthese*, 38, 415–456.

Djurfeldt, M,. Lundqvist, M., Johansson, C., Rehn, M., Ekeberg, Ö., & Lansner, A. (2008). Brain-scale simulation of the neocortex on the IBM Blue Gene/L supercomputer. *IBM Journal of Research and Development*, 52(1/2), 31–41.

Eliasmith, C., Stewart, T. C., Choo, X., Bekolay, T., DeWolf, T., Tang, Y., & Rasmussen, D. (2012). A large-scale model of the functioning brain. *Science*, 338(6111), 1202–1205.

Falit, B. P., & Gross, C. P. (2008). Access to experimental drugs for terminally ill patients. *JAMA: The Journal of the American Medical Association*, 300(23), 2793–2795.

Fukuyama, F. (2002). *Our Posthuman Future*. New York: Farrar Straus & Giroux.

Gamez, D. (2005). An ordinal probability scale for synthetic phenomenology. In Chrisley, R., Clowes, R., & Torrance, S. (Eds.), *Proceedings of the AISB05 Symposium on Next Generation Approaches to Machine Consciousness* (pp. 85–94). Hatfield, Hertfordshire.

Gamez, D. (2008). Progress in machine consciousness. *Consciousness & Cognition*, 17(3), 887–910.

Gold, E. R. (1998). *Body Parts: Property Rights and the Ownership of Human Biological Materials*. Washington, DC: Georgetown University Press.

Goodall, J. (1999). An order of pure decision: Un-natural selection in the work of Stelarc and Orlan. *Body & Society*, 5, 149–170.

Hanson, R. (1994). If uploads come first: The crack of a future dawn. *Extropy*, 6, 10–15.

Hanson, R. (2008). Economics of the singularity. *IEEE Spectrum*, 45(6), 37–42.

Herzog, M. H., Esfeld, M., & Gerstner, W. (2007). Consciousness & the small network argument. *Neural Networks*, 20(9), 1054–1056.

Markram, H. (2006). The blue brain project. *Nature Reviews Neuroscience*, 7, 153–160.

Merkle, R. (1989). *Large Scale Analysis of Neural Structures*. CSL-89-10. Palo Alto, CA: Xerox Palo Alto Research Center.

Metzinger, T. (2003). *Being No One*. Cambridge, MA: MIT Press.

Miller, F. G., & Rosenstein, D. L. (2008). Challenge experiments. In Emanuel, E. J., Grady, C., Crouch, R.A., Lie, R.K., Miller, F.G., Wender, D. (Eds.), *The Oxford Textbook of Clinical Research Ethics* (pp. 273–279). Oxford: Oxford University Press.

Moor, J. H. (1988). Testing robots for qualia. In Otto, H. R., & Tuedio, J. A. (Eds.), *Perspectives on Mind*. Dordrecht, the Netherlands: D. Reidel.

Nordmann, A. (2007). If and then: A critique of speculative nanoethics. *Nanoethics*, 1, 31–46.

Preissl, R., Wong, T. M., Datta, P., Flickner, M. D., Singh, R., Esser, S. K., McQuinn, E. et al. (2012). Compass: A scalable simulator for an architecture for cognitive computing. In *Proceedings of the Supercomputing*. Salt Lake City, UT.

Prinz, J. J. (2003). Level-headed mysterianism and artificial experience. In Holland, O. (Ed.), *Machine Consciousness*. Exeter, Devon: Imprint Academic.

Regan, T. (1983). *The Case for Animal Rights*. Berkeley, CA: The University of California Press.

Regan, T., & Singer, P. (Eds.) (1989). *Animal Rights and Human Obligations* (2nd ed.). Englewood Cliffs, NJ: Prentice Hall.

Sandberg, A. (2013). Feasibility of whole brain emulation. In Müller, V. C. (Ed.), *Theory and Philosophy of Artificial Intelligence* (SAPERE, pp. 251–264). Berlin, Germany: Springer.

Sandberg, A., & Bostrom, N. (2008). *Whole Brain Emulation: A Roadmap*. Oxford: Future of Humanity Institute.

Schofield, J. C. (2002). *Analgesic Best Practice for the Use of Animals in Research and Teaching—An Interpretative International Literature Review*. Wellington, New Zealand: Food and Agriculture Organization of the United Nations (FAO).

Taylor, J. G. (2007). Commentary on the "small network" argument. *Neural Networks*, 20(9), 1059–1060.

Zeigler, P. B. (1985). *Theory of Modelling and Simulation*. Malabar, FL: Krieger.

Zeigler, P. B., Praehofer, H., & Kim, T. (2000). *Theory of Modeling and Simulation: Integrating Discrete Event and Continuous Complex Dynamics Systems*. San Diego, CA: Academic Press.

CHAPTER 11

Long-Term Strategies for Ending Existential Risk from Fast Takeoff

Daniel Dewey

CONTENTS

ABSTRACT   If, at some point in the future, each artificial intelligence (AI) development project carries some amount of existential
risk from fast takeoff, our chances of survival will decay exponentially
until the period of risk is ended. In this chapter, I review strategies
for ending the risk period. It seems that effective strategies will need
to be resilient to government involvement (nationalized projects,
regulation, or restriction), will need to account for the additional difficulty of solving some form of the control problem beyond the mere
development of AI, and will need to deal with the possibility that
many projects will be unable or unwilling to make the investments
required to robustly solve the control problem. Strategies to end the
risk period could take advantage of the capabilities provided by powerful AI, or of the incentives and abilities governments will have to
mitigate fast takeoff risk. Based on these considerations, I find that
four classes of strategy—international coordination, sovereign AI,
AI-empowered project, or other decisive technological advantage—
could plausibly end the period of risk.

11.1 INTRODUCTION

It has been argued that after some level of artificial intelligence (AI) capability is reached, an AI might be able to improve very quickly, and could
gain great enough cognitive capability to become the dominant power on
the Earth.[*] Call this "fast takeoff." In this chapter, I assume that fast takeoff will be possible at some point in the future and try to clarify the resulting strategic situation.

Most work on existential risk (x-risk)[†] from long-term AI capabilities
has focused on the problem of designing an AI that would remain safe
even if it were to undergo a fast takeoff. Bostrom calls this the *control
problem*.[‡]

---

[*] Chalmers, The singularity: A philosophical analysis; Bostrom, *Superintelligence: Paths, Dangers, Strategies*.
[†] Bostrom, Existential risk prevention as global priority.
[‡] *Superintelligence*, p. 128.

Imagine an optimistic future: the control problem has been solved, and a prudent, conscientious project has used the solution to safely develop human-level or even superintelligent AI. The AI race has been won, and the control problem has been solved in time to keep this project from causing harm. Has the danger now passed?

Solving the control problem leaves a major issue: other projects are probably developing AI, each carrying the potential for an existential disaster, and not all of those projects will be as safe as the first one. Some additional strategy is needed to end the period of x-risk from fast takeoff. Furthermore, strategies we could take are probably not equally likely to succeed; maximizing the chances of a positive outcome will require us to choose well among them.

The need for a long-term strategy is not a new insight (see, e.g., Muehlhauser and Bostrom, 2014; Yudkowsky, 2008), but I have not found an overview of strategies for ending AI x-risk, nor much in the way of comparing their strengths and weaknesses (Sotala and Yampolskiy, 2013 comes closest). In this chapter, I attempt such an overview. After introducing the exponential decay model of fast takeoff x-risk (Section 11.2) and reviewing what seem to be the most relevant considerations (Section 11.3), I find that plausible strategies fall into four categories (Section 11.4):

1. International coordination

2. Sovereign AI

3. AI-empowered project

4. Other decisive technological advantage

Implementing one of these strategies may be the best thing one could do to reduce overall x-risk from fast takeoff—in fact, if the considerations underlying my analysis are correct, then it seems plausible that x-risk from fast takeoff cannot be mitigated significantly without using one of these strategies. Based on this analysis, projects aiming to reduce fast takeoff x-risk should be aiming to eventually implement one of these strategies, or to enable future projects to implement one of them.

11.2 EXPONENTIAL DECAY MODEL OF FAST TAKEOFF X-RISK

For the rest of the chapter, I will assume that at some point in the future, an AI will be able to improve quickly enough to become the dominant power on the Earth (gaining what Bostrom calls "decisive strategic advantage"'"). There is disagreement about this assumption that I cannot hope to settle decisively here. Some of the best places to pick up the thread of this debate in the literature include: (Bostrom, 2014; Hanson, 2014; Hanson and Yudkowsky, 2013; Yudkowsky, 2013).

Notably, this assumption means that my analysis will not apply to futures in which no one agent can gain a decisive advantage and gains from AI progress are consequently spread over a large number of human, AI, and hybrid agents with roughly comparable abilities and rates of growth (as suggested in, e.g., Hanson, 2014; Kurzweil, 2005, p. 301). The possibility of a fast takeoff also makes it much less likely that there will be many AIs at roughly similar levels of capability, because this would require all of those AIs to begin their takeoffs very close together in time.[†]

If an AI has gained a decisive strategic advantage, and if it is configured to choose actions that best bring about some consequences (its "goal") and it is not under some kind of human control, the AI might find that eliminating humans altogether is the best course of action to reliably achieve its goal,[‡] or it might use up the resources that we need for a thriving future. Either of these would be an existential catastrophe.

Given this, if after some future time each AI project carries some nonnegligible independent chance of creating such an AI, the most important features of our situation can be described with what I call the "exponential decay model" of fast takeoff x-risk:

> *Exponential decay model*: A long series of AI projects will be launched and run, each carrying nonnegligible independent x-risk. Even if each project carries a small chance of catastrophic failure, as more and more teams try their luck, our chances of survival will fall very low.

The exponential decay model may not be correct: for example, it may be that one project develops such cheap and advanced AI that nobody else is motivated to start new projects, or path dependencies set all projects on a common course with low independent risk of disaster, or one project develops techniques that cheaply lower the risk incurred by all other projects to a negligible level. However, I think (motivated in part by the

---

[*] *Superintelligence*, p. 78.

[†] *Superintelligence*, p. 82.

[‡] If an AI were to have a decisive strategic advantage, why would eliminating humans be the best way to bring about its goals? How can we threaten its goal achievement? There are at least two possibilities: removing humans from the picture may be how the AI gets decisive strategic advantage, or its decisive strategic advantage may consist largely in its ability to remove humans from the game altogether.

considerations in Section 11.3) that this model is plausible enough that it probably accounts for a large proportion of all x-risk from fast takeoff.

There are many questions we could ask about the exponential decay model, but I will focus on strategies that reduce fast takeoff x-risk by ending the risk period. This would require that (1) any subsequent projects that are started must be negligibly risky and (2) all existing projects must be halted or their risk must be reduced to an acceptable level. Which strategies could be used to accomplish this depends on how hard it is to stop projects or render them safe, and on what types of influence are available.

11.3 MAJOR CONSIDERATIONS

Three considerations seem most important to which strategies could plausibly end the risk period:

1. *Government involvement*: Government involvement with AI development, either through regulation or through nationalized projects, is reasonably likely, especially if the risk period continues for a reasonably long time. Nationalized projects will be unusually difficult to halt or prevent, but governments will be able (and may be motivated) to bring significant influence to bear on the problem of fast takeoff x-risk.

2. *The control problem*: If the problem of using superintelligent AI safely, that is, without incurring widespread and undesired side effects, is as difficult as it appears, then raising the safety level of all projects to acceptable levels will be very difficult (especially given the large number of projects that could be started if the risk period continued indefinitely, and given how widely different projects' beliefs about AI risk would plausibly vary), and using AI as a source of influence to solve the problem would carry significant technical challenges.

3. *Potential for harnessing fast takeoff*: Fast takeoff is the main source of x-risk from AI, but if some form of the control problem can be solved, then it could be a very powerful and useful tool for reducing overall x-risk from fast takeoff.

11.3.1 Government Involvement

Governments may come to believe that AI is a significant factor in national security and/or prosperity, and as a result, governments may sponsor AI projects or take steps to influence AI development.* Efforts to reduce unsafe AI development, whether by making projects safer or by halting unsafe ones, will need to take government incentives and actions into account.

- *Extreme incentives*: Governments may view advanced AI as an existential threat, because there may be no good defense against very powerful AI, undermining a nation's abilities to protect even its domestic interests. This could result in nationalized AI projects with strong mandates, or in strong government incentives to regulate and coordinate with other governments.

- *Nationalized projects*: Nationalized projects could have very different characteristics than commercial projects: they would have unusual powers at their disposal (law, espionage, police, or military force), would not be easily dissuaded through commercial incentives, might enjoy a legal monopoly within some jurisdiction, and might be more easily influenced to adopt safety and control policies.

- *Legal powers*: Governments may regulate, restrict, or monitor AI research within their jurisdictions. There could also be international agreements about how AI projects may or may not be conducted.

- *Broad cooperation*: Governments may be more apt than commercial projects to cooperate and coordinate globally and to promote some form of public good.

It is not clear what degree of government involvement we should expect. Perhaps it is unrealistic to think that governments will attend to such unlikely sounding risks, or perhaps it is unrealistic to expect any AI project to advance far without being nationalized.† It may be that countries do not decide to create nationalized AI projects; governments' present-day reliance on foreign computer hardware and software could set a precedent for this. Even given this uncertainty, however, it seems that if the period of fast takeoff x-risk continues for some time, the probability of government involvement will certainly increase and may increase significantly.

Strategies that I consider reasonable do account for the possibility of government involvement, especially if the risk period lasts for some time;

---

* *Superintelligence*, p. 78.
† *Superintelligence*, p. 85.

this is why, for instance, purely commercial incentives are probably not sufficient to suppress unsafe AI projects. I also think it is reasonable to consider strategies that make use of government influence to end the period of fast takeoff risk, although it is not clear how easy or hard it will be to bring that influence to bear successfully.

11.3.2 Control Problem

Bostrom defines "the control problem" as "the problem that a project faces when it seeks to ensure that the superintelligence it is building will not harm the project's interests."* I will emphasize the difficulty of using super-intelligent AI without incurring widespread and unwanted side effects.

The control problem has two major implications for risk-ending strategies: first, if the control problem is hard, it will be difficult to improve many AI projects to an acceptable level of safety without monitoring and restricting them fairly closely; and second, technical work on some form of the control problem will probably be necessary if powerful AI is to be used to end the risk period.

Why would we expect the control problem to be hard? I will summarize some reasons here, though much of the AI risk literature has been devoted to this question. *Superintelligence* covers most of these aspects in much greater detail:

- *High impact*: In order to keep the level of expected risk acceptably low, solutions to the control problem will need to have very low chances of failure, in both the objective sense (high reliability in almost all circumstances) and the subjective sense (justifiably high confidence that our safety models and assessments are correct). A project's level of reliability would need to be much higher than it is typically in commercial or academic software development.

- *Diversity of options and difficulty of thorough goal specification*: A superintelligent AI would have a great diversity of options and plans available to it. Among these plans would be many that have what we would regard as highly undesirable side effects. A full set of goal criteria that avoid any undesirable side effects seems difficult to achieve. It could also be very difficult to fully verify a plan once it has been produced. Though the project would have the use of a

---

* *Superintelligence*, p. 128.

superintelligent AI for this task, there is a chicken-and-egg problem: if, by assumption, there is some criterion that was not used by the AI when producing the plan, it is not clear how the AI could help the project notice that the produced plan fails that same omitted criterion. A full examination of this issue is beyond the scope of this chapter; more can be found especially in *Superintelligence*, chapters 6, 8, 9, and 12.

- *Technical reliability failures*: Even if goal specification problems were solved, there are technical reliability problems that could arise as an AI becomes much more intelligent, especially if it is undertaking a series of self-improvements. For example, an AI could fail to retain its safety properties under self-improvement, or could go systematically wrong through failures of reasoning about logical uncertainty and certain types of decision theoretic problems. The research output of the Machine Intelligence Research Institute[*] is the best source for information about these kinds of technical reliability failures.

Beyond these aspects of the control problem, there are a handful of considerations pertaining to the control problem's role in the exponential decay model as a whole:

- *Different beliefs about the control problem among projects*: Different projects may make differing assessments of the necessity and difficulty of solving the control problem. This could lead to a "unilateralist's curse,"[†] in which the least cautious project triggers a globally bad outcome. Additionally, some projects will probably have a safety standard closer to standard academic or commercial software engineering than safety-critical systems engineering.

- *Safety/speed trade-offs*: Projects will need to split their resources between the control problem and AI development. Additionally, developers could spend arbitrary amounts of time checking the safety of each stage of their projects. These are two minimal ways that speed and safety of development could be traded off against one another, but there could be others.[‡]

---

[*] http://intelligence.org/research/

[†] Bostrom, Sandberg, and Douglas, *The Unilateralist's Curse*.

[‡] *Superintelligence*, p. 246; Armstrong, Bostrom, and Shulman, *Racing to the Precipice: A Model of Artificial Intelligence Development*.

How could these considerations fail to hold? It might be that the control problem is significantly easier than it now appears, or that progress in AI will clear up these difficulties before significant damage can occur. It might also be that some project will be able to largely solve the control problem, and then to communicate their solution in a way that can be cheaply and easily implemented by all other AI projects. Overall, however, it seems reasonable to me to require a strategy to end the risk period to cope adequately with the apparent difficulty of the control problem.

11.3.3 Potential for Harnessing Fast Takeoff

If an AI could undergo fast takeoff, then it may also be that an AI project could (by solving some aspects of the control problem) gain access to great technological development capabilities, superintelligent strategic ability, cognitive labor that scales with hardware, and any number of the other "superpowers" Bostrom enumerates.* Though it appears to be quite difficult to end the fast takeoff risk period, these capabilities seem powerful enough to be up to the task.

Particularly useful capabilities seem to be strategic planning, technological development, and highly efficient inference (for data mining and surveillance tasks). However, if the control problem is prohibitively difficult to solve, or it cannot be solved satisfactorily before many other projects have incurred significant fast takeoff risk, then the potential for harnessing fast takeoff will not be all that valuable.

11.4 PLAUSIBLE STRATEGIES FOR ENDING FAST TAKEOFF RISK

Given these considerations, effective strategies will need to be resilient to government involvement (nationalized projects, regulation, or restriction, especially as the period of risk goes on), will need to account for the additional difficulty of solving some form of the control problem beyond the mere development of AI, and will need to deal with the possibility that many projects will be unable or unwilling to make the investments required to robustly solve the control problem. Strategies could take advantage of the capabilities provided by powerful AI, or of the incentives and abilities governments will have to mitigate fast takeoff risk.

In this section, I will describe four types of strategy that seem to meet these requirements. Some of these strategies put in place defenses which

---

* *Superintelligence*, p. 91.

could degrade or fail altogether over time; for example, a treaty is subject to future political change. If this happens, a transition will have to be made to another strategy for preventing fast takeoff risk.

As a reminder, I have assumed that fast takeoff will eventually be possible, that the exponential decay risk model is reasonably accurate, and that it will eventually be necessary to somehow end the period of risk. If these assumptions are not all true, then it may be that none of these strategies are necessary, and some other course of action is best.

11.4.1 International Coordination

At around the time the risk period would begin, a large enough number of world governments could coordinate to prevent unsafe AI development within their areas of influence. This could take the form of an AI development convention, that is, a set of strict safety rules for certain kinds of AI development. Alternatively, international coordination could be used to create a joint international AI project with a monopoly on hazardous AI development, perhaps modeled after the Baruch Plan, a 1946 nuclear arms control proposal. Based on the Acheson–Lilienthal Report,* the Baruch Plan called for the creation of an International Atomic Development Authority, a joint international project that would become "the world's leader in the field of atomic knowledge" and would have the "power to control, inspect, and license all other atomic activities."† A Baruch-like plan would create an International Artificial Intelligence Authority with similar powers and responsibilities, and any potentially hazardous AI research and use would be conducted by this group (as Bostrom suggests might succeed‡). In the Baruch Plan, the Authority was to "supplement its legal authority with the great power inherent in possession of leadership in [atomic] knowledge," that is, it was to use its nuclear monopoly to ensure that the terms of the Plan were not violated by other countries. Similarly, an AI Authority might optionally be an AI-empowered project (described later in this section), using its AI advantage to ensure that the terms of the coordination are not violated.

The Acheson–Lilienthal Report and later the Baruch Plan were formulated specifically to fit with the technical facts about how nuclear weapons could be built, how peaceful nuclear power could be pursued, how those

---

* Atomic Energy and Lilienthal, *A Report on the International Control of Atomic Energy.*
† Baruch, *The Baruch Plan.*
‡ *Superintelligence,* p. 86, p. 253.

two processes overlapped and diverged, and what kinds of materials and techniques played key roles. For example, the creation of an Authority was inspired in part by the fact that only particular phases of the development of atomic energy were "potentially [or] intrinsically dangerous," so that assigning these phases to the Authority would offer good assurance of safety.* Analogously, it is plausible that further technical insight into fast takeoff risk might suggest other forms of international cooperation that are more suitable to this particular problem.

The characteristic challenge of international coordination strategies is political in nature. Failures of coordination could lead to failure to create a convention or Authority at all, or failure to enforce the agreed-upon terms well enough to prevent existential disasters.

11.4.1.1 Benefits

- x-risk prevention is a global public good, and thus, making it the subject of international coordination is natural. x-risk from fast takeoff has a strong unilateralist component, and reducing it may require the use of powers normally reserved for governments, so this strategy is in some ways the most common sense way to solve the problem. This matters because more sensible-sounding strategies are easier to explain and gather support for, and may also (if common sense is reliable in this scenario) be more likely to succeed. For similar reasons, it is useful that this strategy fits with democratic ideals.

- An internationally coordinated project would plausibly be in a position to corner the market on many resources needed for AI work—highly skilled researchers, computing power, neuroscience hardware, data sets, and so on—and thus it might have a good chance of outracing any illegal "rogue" projects. (However, such a project might look to the historical example of the public Human Genome Project and the private Celera project for ways that a public project can fail to corner the market on public good science and engineering challenges.)

- If we are quite confident that only the joint AI project is running, then the race dynamic disappears entirely, leaving us as much time as needed to solve the control problem (if this is desired).

---

* Atomic Energy and Lilienthal, *A Report on the International Control of Atomic Energy*, p. 26.

11.4.1.2 Difficulties

- Preventing secret nationalized or private projects could be very difficult and would require unprecedented transparency, inspection, and cooperation, as well as expanded powers of surveillance and intervention over private enterprise.

- These strategies would require a very strong consensus on which kinds of AI development are hazardous, both for logistical reasons and in order to justify the high costs of preventing development and verifying cooperation. This consensus could fail because of the difficulty or cross-disciplinary nature of fast takeoff risk questions, the unavailability of relevant information, or motivated manipulation of the scientific and political communities by groups that believe they would benefit from preventing AI regulation.

- The implementation of a convention or Baruch-like plan could fail for political reasons. Though the exact reasons the Baruch Plan failed are hard to know for certain, three common proposals are that the Soviets suspected that they would be outvoted by the United States and its allies in the governing body of the joint project, that the Soviets believed their sovereignty would be compromised by inspections, and that the United States would keep its nuclear monopoly for some time as the plan was implemented.[*] Analogous political difficulties could easily arise in the case of AI, but might be mitigated somewhat if the plan was developed before some parties could "pull ahead" in the race to AI, as Bostrom suggests.[†]

11.4.1.3 Failure

If this strategy fails because political consensus cannot be reached, then it seems that it will fail without blocking other strategies too badly. If this strategy fails after it is implemented, this failure may not be recoverable: a joint project could fail to be safe and trigger an existential disaster, or it could prevent conscientious projects from proceeding although reckless projects continue.

---

[*] Russell, *Has Man a Future?*; Wittner, *One World or None: A History of the World Nuclear Disarmament Movement through 1953.*
[†] *Superintelligence*, p. 253.

11.4.2 Sovereign AI

In a sovereign AI strategy, a private or government-run project creates an autonomous AI system, which in turn acts to end the period of risk. Bostrom defines a sovereign AI as a "system that has an open-ended mandate to operate in the world in pursuit of broad and possibly very long-range objectives."* Here, I mean "sovereign" in Bostrom's sense, and not in the sense of a ruler or monarch; though an AI with a decisive strategic advantage would have the capability required to become a sovereign in this more traditional sense, it need not be designed to exert widespread or invasive control.

Why would we think that AI itself could be a useful tool for mitigating fast takeoff risk? As I suggested earlier (and as has been pointed out by many in the AI x-risk literature, e.g., Bostrom, 2014; Muehlhauser and Bostrom, 2014; Yudkowsky, 2008), if a fast takeoff is possible, then a carefully designed AI might be able to undergo fast takeoff, granting it or its project capabilities that could be used to end the risk period. These capabilities would plausibly not be accessible in any other way, at least on similar timescales. These advantages could be used by a sovereign AI, or (see Section 11.4.3) by an AI-empowered project.

In order to end the risk period, a sovereign AI would need to either prevent unsafe fast takeoffs or prevent them from being extinction risks in some other way. In thinking about ways to do this, I have found it useful to make an informal distinction between two types of sovereign AI: proactive and reactive.

A proactive sovereign AI is designed to end the period of risk by intervening significantly in the world, altering one or more of the basic conditions required for a fast takeoff into an existential catastrophe. After its takeoff, a proactive sovereign AI might (covertly or overtly) prevent AI development from proceeding beyond a certain point, might make sure that any projects that do continue are safe, or might merely accrue resources and build institutions sufficient to prevent any AI that does undergo fast takeoff from causing significant harm. Useful capabilities for these purposes would include strategic planning, persuasion, technological development, hacking, and economic productivity. A proactive sovereign AI could also be tasked with broader fulfilment of humane values,† or it could be designed to only intervene only on AI development (something like Goertzel's "Nanny AI" proposal‡).

---

* *Superintelligence*, p. 148.
† Yudkowsky, Complex value systems in friendly AI.
‡ Goertzel, Should humanity build a global AI nanny to delay the singularity until it's better understood?

A reactive sovereign AI is designed to end the risk period by preparing to respond to a fast takeoff in a way that halts it, contains it, or alters it to be harmless. Instead of intervening significantly in the world before a fast takeoff, it acts as a "just-in-time" response to AI catastrophes, thus ensuring that all AI development is "safe" (because no disaster will actually proceed to completion). For example, after its takeoff, a reactive sovereign AI may use its capabilities to set up a global surveillance system and wait for an uncontrolled takeoff to begin, then intervene to halt the offending program. Depending on how effectively a reactive sovereign AI can infer states of the world from limited information, deploying a new surveillance system may not be necessary at all; perhaps it will be sufficient for the AI to use the existing Internet, cell phone networks, and news channels. Because a reactive sovereign is designed to wait until a disaster is under way, its intervention options are more limited; it will probably need to be able to directly interact with the hardware involved in an ongoing intelligence explosion.

Sovereign AI strategies seem to require very robust solutions to most aspects of the control problem, because autonomous superintelligent AI will probably not be under the direct control of its developers for monitoring and correction, and its actions will not be mediated by humans. Reactive sovereign AI strategy may put more focus on "domesticity" of motives, in which the general idea is that instead of being motivated to manage fast takeoff risk in whatever way is most effective, the reactive sovereign AI would be motivated to prevent certain events without having too many or too large impacts on other aspects of the world.* It is not clear how technically difficult this would be.

11.4.2.1 Benefits

- Sovereign AI strategies could be undertaken by an international, national, or private AI project, assuming that they are sufficiently far ahead in their AI capabilities.

- Sovereign AI strategies do not require a broad scientific or political consensus; if it is extremely difficult to convincingly communicate fast takeoff risk considerations or control problem solutions, it could be a significant advantage of only having to convey these considerations to a single project instead of to a broader community.

---

* *Superintelligence*, p. 140.

- If well engineered, a sovereign AI could be considerably more responsive, able, and reliable than human governments or projects attempting similar tasks.

11.4.2.2 Difficulties

- As noted in "failure," a project that aims to create a sovereign AI is taking on a very large responsibility, because failure in a variety of directions could be an existential disaster. It would be an unprecedentedly important and dangerous project.

- Solving the control problem well enough to be confident that launching a sovereign AI has positive expected value might be prohibitively difficult.

- Sovereign AI strategies require a project to have a lead large enough that they can solve whatever parts of the control problem are relevant to their strategy and then implement their safety plan before any other projects can trigger an existential disaster. Given the difficulty of these problems, this lead may have to be pretty large, although it is not clear how to quantify this.

- Once a sovereign AI is active and gains a significant strategic advantage, it could become very difficult to deactivate or repurpose it (because many goals will be worse fulfilled if the AI allows itself to be deactivated or repurposed). Unless some aspects of the control problem are solved, switching from a sovereign AI plan to some other plan might not be possible after a certain point.

- If a sovereign AI is tasked with a broader fulfilment of humane values, setting it loose is a very large bet; if the AI is expected to gain a decisive strategic advantage, it could well control the majority of future resources available to us and be solely responsible for how well humanity's future goes.

- Creation of a sovereign AI with the potential for decisive strategic advantage would probably significantly decrease the powers of current institutions; if these institutions are not involved, or do not think their interests will be satisfied, they may oppose development.

11.4.2.3 Failure

- AI with a decisive strategic advantage is the main source of x-risk from artificial intelligence; creating a sovereign AI is inherently risky, and creating an inhumane sovereign by accident would be an existential catastrophe.

- A sovereign AI project could fail by simply losing the race to AI to another project. This could happen because the project has fewer resources, which are trying to solve harder or more problems, or if its technology is leaked or stolen by competing projects.

- If projects and AIs are designed to shut down cleanly when sufficient assurance of a well-made sovereign cannot be reached, then switching to another strategy seems possible. Existence of AI projects may make international coordination harder. It may or may not be that the failed sovereign project can be easily repurposed into a reactive sovereign or into an AI-empowered project strategy; which is more likely is not clear.

11.4.2.4 Aside: Proactive vs. Reactive Strategies

The proactive/reactive distinction applies to both sovereign AI strategies and AI-empowered project strategies. Before I move on to AI-empowered project strategies, I will discuss some of the differences between proactive and reactive strategies in general.

- Some proactive strategies might seem to be overstepping the rights or responsibilities of the projects executing them (private projects, individual governments, or incomplete international coalitions). Proactive strategies may seem distasteful, paternalistic, or violent, and they could carry significant costs in terms of enforcement, option restriction, or chilling effects. Historically, although it was proposed that the United States could take a proactive strategy and use its temporary nuclear monopoly to gain a permanent one by threatening strikes against countries that started competing programs,[*] this strategy was not ultimately taken; the same or similar motivations might prevent projects or governments from pursuing proactive AI strategies. For these reasons, governments seem to be more likely to favor reactive strategies over proactive ones.

---

[*] Griffin, *The Selected Letters of Bertrand Russell: The Public Years, 1914–1970, Bd 2.*

- It seems that reactive strategies are, by definition, leaving some opportunities to mitigate x-risk on the table, for example, by refraining from stopping high-risk AI development projects early. It is not clear how low this residual risk could be driven by a reactive AI's decisive strategic advantage.

- Reactive strategies do not "end the AI race" in the way that proactive strategies do. If projects continue to push the frontier of accident-free AI development, a reactive sovereign or AI-empowered project may have to continue improving itself at a rate fast enough to maintain its decisive strategic advantage. It is not clear how difficult this will be; it seems that the viability of reactive strategies in the long run will depend on the overall shape of AI and other technological development curves.

11.4.3 AI-Empowered Project

By "AI-empowered project," I mean a project that uses some nonsovereign AI system(s) to gain a decisive strategic advantage. Like the sovereign AI strategy, AI-empowered projects could be proactive or reactive.

As in the sovereign AI strategy, a proactive AI-empowered project would need to gain enough influence (of some sort) to prevent unsafe fast takeoffs or prevent them from being extinction risks in some other way. A project could use any combination of Bostrom's superpowers to do this. Perhaps, for example, they could use exploits to monitor and alter activity on any computer with a standard Internet connection; they could implement a superintelligence-based persuasion campaign to gain whatever type of influence they need to reliably make all other AI projects safe, or end them; or they could use their technological development abilities to deploy a large-scale surveillance and intervention infrastructure. It is difficult to categorize all of the approaches an AI-empowered project could use, because it is not yet clear exactly what abilities they will have or what environment they will be working in, but a more in-depth study could shed some light on the choices an AI-empowered project would have.

A reactive AI-empowered project would maintain its decisive strategic advantage while intervening only to prevent unsafe fast takeoffs. As mentioned above, this reactive strategy does not end the AI race, and the AI-empowered project may need to devote some of its resources to keeping a large enough lead over other projects. In theory, a reactive project might be extremely covert, using its advantage to maintain widespread and thorough surveillance with the potential for precise interventions if an existential disaster should begin.

To succeed with an AI-empowered project strategy, a project might not need to solve the problem of specifying fully humane-valued motivation, but they would need to solve some form of the control problem—perhaps domesticity, safe intelligence explosion, and whatever is needed to make a safe "oracle" or "tool" AI,* which can be used to answer questions and plan strategies (technical or political). They would also need to solve organizational problems, maintaining very reliable control of the considerable power at their disposal in the face of external (and possible internal) opposition; it is possible that AI tools could be helpful for these purposes.

11.4.3.1 Benefits

- Empowered projects could be international, national, or private, and would not require a broad consensus.

- As long as the empowered project cooperates, it seems possible to transition to any other strategy.

- An AI-empowered project might not need to solve some parts of the control problem that a sovereign AI project would need to solve, and might also be more robust against unexpected AI behaviors and malfunctions than a sovereign AI project would be, because their AI systems might be localized and have more limited capabilities.

11.4.3.2 Difficulties

- Like sovereign AI strategies, AI-empowered project strategies require an AI project to have a large enough lead on other AI projects.

- A sovereign AI could be considerably more capable than humans attempting similar tasks, even if those humans are using AI tools. A minimal example of this is reaction time: even if an AI-empowered project is using some kind of direct brain–machine interface, the speed at which a human can react and make decisions in an emergency situation will be much slower than the speed at which a sovereign AI can react. This problem will be worse if an empowered project wishes to work and makes decisions as a team.

---

* *Superintelligence*, p. 145.

- An AI-empowered project would face serious organizational challenges and would need to be robust against internal or external opposition.

11.4.3.3 Failure

- An AI-empowered project could accidentally create a sovereign AI, or otherwise lose control of their AI, which could trigger an existential disaster.

- As above, an AI-empowered project could fail by simply losing the race to AI to another project, perhaps through leaks.

- An AI-empowered project would have an unprecedented amount of control over global affairs, and there could be significant risk of it abusing this control; the historical precedent for humans behaving well when they suddenly come into large amounts of political power does not seem particularly encouraging.

11.4.4 Other Decisive Technological Advantage

The development of atomically precise manufacturing, whole brain emulations, or some other comparably revolutionary technology could enable a private project or government(s) to gain a decisive strategic advantage, and then to enact a proactive or reactive empowered project strategy. Whatever advantage was found would have to be powerful enough either to allow the empowered project to find and halt unsafe AI development globally, or to detect and prevent AI catastrophes as they begin. Perhaps whole brain emulation[*] or atomically precise manufacturing[†] would be sufficient for this purpose, or perhaps further comparable sources of advantage have yet to be discovered.

11.5 CONCLUSION

Starting from the assumption that an AI could improve rapidly enough to gain a decisive strategic advantage, I have argued that the exponential decay model of fast takeoff x-risk captures the most important features of the situation. After describing the most relevant considerations, I have reviewed the four strategies that seem that they could plausibly end the period of risk: international coordination, sovereign AI, AI-empowered project, and other decisive technological advantage.

---

[*] Sandberg and Bostrom, *Whole Brain Emulation: A Roadmap.*
[†] Drexler, *Radical Abundance: How a Revolution in Nanotechnology Will Change Civilization.*

If this analysis is correct, then these strategies can give significant guidance to projects aiming to mitigate AI x-risk. If the majority of AI x-risk comes from fast takeoff, then risk reduction projects should be aiming to eventually implement one of these strategies, or to enable future projects to implement one of them. Additionally, the choice of which strategy or strategies a project ought to pursue seems to be a very important one.

It also seems to me that the mitigation strategies are far enough from business as usual that it is probably not feasible to implement them as a hedge against the mere possibility of fast takeoff x-risk; international coordination, or even the level of support necessary to create sovereign AI or an AI-empowered project, may not be feasible in the face of considerable uncertainty about whether an AI could improve rapidly enough to gain a decisive strategic advantage. Becoming more certain about whether or not AI can realistically gain a decisive strategic advantage—whether fast takeoff is possible—would therefore be a valuable pursuit for a risk mitigation project.

ACKNOWLEDGMENTS

I thank Nick Beckstead, Paul Christiano, Owen Cotton-Barratt, Patrick LaVictoire, and Toby Ord for their helpful discussion and comments.

APPENDIX: FURTHER STRATEGIC QUESTIONS

Though I think these considerations and strategies do a fair amount to map out the space, there are many aspects of the treatment that are clearly incomplete, and there is still room for major changes and updates— perhaps it will even turn out that fast takeoff is impossible or unlikely, that the exponential decay model is not very accurate, or that ending the risk period is not the best way to reduce overall x-risk from AI.

When selecting questions to answer next, it is important to consider the timing of labor to mitigate x-risk.[*] It does seem plausible to me that answers to strategic questions could lead to significant course changes or could be helpful in increasing the resources (human, material, or attentional) devoted to superintelligent AI risk, but if they were not likely to, or if their answers are only valuable in very particular scenarios that we may not encounter, then it may be more useful to do other kinds of work.

---

[*] Ord, *The Timing of Labour Aimed at Reducing Existential Risk.*

That said, there are three strategic questions seem most important to me at this stage:

- What fraction of x-risk from AI comes from fast takeoff?

- How much credence should we give to the key premises of fast takeoff risk, the exponential decay model, and the major strategic considerations?

- Assuming the exponential decay model holds, is it the case that ending the period of risk is the best aim, or are there other ways to reduce overall fast takeoff risk more effectively?

There are many other questions that could help us choose among these strategies, or that could give us more insight into how each strategy could play out:

- How do the exponential decay model, major considerations, and strategies change if human-level AI comes sooner (<20 years) or later?

- Suppose that we thought one of these strategies was best, or that we had some distribution over which was going to be best. What actions could we take now to best improve our chances of future success?

- What evidence will likely be available about these things as we approach and enter the period of risk, and how persuasive will it be to various parties?

- How hard will it be to solve different aspects of the control problem? How do these difficulties affect sovereign, empowered project, proactive, and reactive strategies?

- Can international cooperation succeed? What kinds? Are there historical analogs, and how often have they been successful?

- What organizational challenges would different types of AI-empowered projects face? How difficult will it be to overcome these challenges?

- What abilities could an AI-empowered project plausibly have? Will it be possible for them to gain the strategic advantage that they would need to mount a successful proactive or reactive strategy? Can we make more detailed plans for AI-empowered projects ahead of time?

BIBLIOGRAPHY

Armstrong, Stuart, Nick Bostrom, and Carl Shulman. *Racing to the Precipice: A Model of Artificial Intelligence Development.* Future of Humanity Institute Technical Report 1, Oxford University, 2013.

Baruch, Bernard. *The Baruch Plan.* Presentation to the United Nations Atomic Energy Commission, New York, 1946, June 14.

Bostrom, Nick. Existential risk prevention as global priority. *Global Policy* 4(1) (2013), 15–31.

Bostrom, Nick. *Superintelligence: Paths, Dangers, Strategies.* Oxford University Press, Oxford, 2014.

Bostrom, Nick, Anders Sandberg, and Tom Douglas. *The Unilateralist's Curse.* Future of Humanity Institute, Oxford University.

Chalmers, David. The singularity: A philosophical analysis. *Journal of Consciousness Studies* 17(9–10) (2010), 7–65.

Drexler, K. Eric. *Radical Abundance: How a Revolution in Nanotechnology Will Change Civilization.* Public Affairs, New York, 2013.

Goertzel, Ben. Should humanity build a global AI nanny to delay the singularity until it's better understood? *Journal of Consciousness Studies* 19(1–2) (2012), 1–2.

Griffin, N. *The Selected Letters of Bertrand Russell: The Public Years, 1914–1970, Bd 2.* Houghton Mifflin, Boston, MA, 2001.

Hanson, Robin. *I Still Don't Get Foom.* Blog post. 2014. http://www.overcoming-bias.com/2014/07/30855.html.

Hanson, Robin, and Eliezer Yudkowsky. *The Hanson-Yudkowsky AI-Foom Debate.* 2013. http://intelligence.org/ai-foom-debate/.

Kurzweil, Ray. *The Singularity Is Near: When Humans Transcend Biology.* Penguin, New York, 2005.

Muehlhauser, Luke, and Nick Bostrom. Why we need friendly AI. *Think* 13(36) (2014), 41–47.

Ord, Toby. *The Timing of Labour Aimed at Reducing Existential Risk.* Blog post. 2014. http://www.fhi.ox.ac.uk/the-timing-of-labour-aimed-at-reducing-existential-risk/.

Russell, Bertrand. *Has Man a Future?* Greenwood Press, New York, 1984.

Sandberg, Anders, and Nick Bostrom. *Whole Brain Emulation: A Roadmap.* Future of Humanity Institute Technical Report 3, Oxford University, 2008.

Sotala, Kaj, and Roman Yampolskiy. *Responses to Catastrophic AGI Risk: A Survey.* Machine Intelligence Research Institute Technical Report. 2013. http://intelligence.org/files/ResponsesAGIRisk.pdf.

United States. Department of State. Committee on Atomic Energy, and David E. Lilienthal. *A Report on the International Control of Atomic Energy.* U.S. Government Printing Office, Washington, DC, 1946.

Wittner, Lawrence S. *One World or None: A History of the World Nuclear Disarmament Movement through 1953.* Vol. 1. Stanford University Press, New York, 1993.

Yudkowsky, Eliezer. Artificial intelligence as a positive and negative factor in global risk. *Global Catastrophic Risks* 1 (2008), 303.

Yudkowsky, Eliezer. Complex value systems in friendly AI. *Artificial General Intelligence.* Springer, Berlin, 2011, pp. 388–393.

Yudkowsky, Eliezer. *Intelligence Explosion Microeconomics.* Machine Intelligence Research Institute Technical Report. 2013. http://intelligence.org/files/IEM.pdf.

CHAPTER 12

Singularity, or How I Learned to Stop Worrying and Love Artificial Intelligence

J. Mark Bishop

CONTENTS

ABSTRACT   Professor Stephen Hawking has recently warned about the growing power of artificial intelligence (AI) to imbue robots with the ability to both replicate themselves and increase the rate at which they get smarter—leading to a tipping point or "technological singularity" when they can outsmart humans. In this chapter, I will argue that Hawking is essentially correct to flag up an existential danger surrounding widespread deployment of "autonomous machines," but wrong to be so concerned about the singularity, wherein advances in AI effectively make the human race redundant; in my world, AI—with humans in the loop—may yet be a force for good.

12.1 BACKGROUND: THE "TECHNOLOGICAL" SINGULARITY

It is not often that you are obliged to proclaim a much-loved international genius wrong, but in his alarming prediction regarding artificial intelligence (AI) and the future of humankind, I believe Professor Stephen Hawking is. Well, to be precise, being a theoretical physicist—in an echo of Schrodinger's cat, famously both *dead and alive* at the same time—I believe the eminent Professor is both *wrong and right* at the same time.* *Wrong* because there are strong grounds for believing that computers will never be able to replicate all human cognitive faculties and *right* because even such emasculated machines may still pose a threat to mankind's future existence; an existential threat, so to speak.

In a television interview on December 2, 2014, Rory Cellan-Jones asked how far engineers had come along the path toward creating AI to which, slightly alarmingly, Professor Hawking replied "Once humans develop artificial intelligence it would take off on its own and redesign itself at an ever increasing rate. Humans, who are limited by slow biological evolution, couldn't compete, and would be superseded."

Although warranting headlines that week, such predictions are not new in the world of science and science fiction; indeed my ex-colleague at the University of Reading, Professor Kevin Warwick, made a very similar prediction back in 1997 in his book *March of the Machines*. In the book, Kevin observed that, even in 1997 there were already robots with the "brain power of an insect"; soon, he predicted that there would be robots with the brain power of a cat, and soon after that there would be machines as intelligent as humans. When this happens, Warwick predicted that the science fiction nightmare of a "terminator" machine could quickly become reality, because these robots will rapidly become more intelligent and superior in their practical skills than the humans that designed and constructed them.

The notion of the singularity (with the accompanying vision of a future mankind subjugated by evil machines) is based on the ideology that *all* aspects of human mentality will eventually be instantiated by an AI

---

* This chapter extends a brief essay first published in *Scientia Salon*, March 2015.

program running on a suitable computer— so-called strong AI.* Of course, *if* this is possible, accelerating progress in AI technologies—caused by both the use of AI systems to design ever more sophisticated AIs and the continued doubling of raw computational power every 2 years as predicted by Moore's law—will eventually cause a runaway effect wherein the AI will inexorably come to exceed human performance *on all tasks*—the so-called point of [technological] "singularity" popularized by the Google futurologist Ray Kurzweil [1].

And at the point, this singularity occurs, so Warwick, Kurzweil, and Hawking suggest that humanity will have effectively been "superseded" on the evolutionary ladder and will be obliged to eek out its autumn days listening to Pink Floyd and gardening—or in some of Hollywood's more dystopian visions, cruelly subjugated or exterminated by "terminator" machines.

I did not endorse these concerns in 1997 and do not do so now; although I do share—for very different and mundane reasons that I will outline later—the worry that AI potentially poses a serious risk to humanity.

12.2 HUMANITY GAP

There are many reasons why I am skeptical of grand claims made for future computational AI, not least empirical. This history of the subject is littered with researchers who have claimed a breakthrough in AI as a result of their research, only for it later to be judged harshly against the weight of society's expectations. All too often these provide examples of what Hubert Dreyfus calls "the first-step fallacy" [2]—undoubtedly climbing a tree takes a monkey a little nearer the moon, but tree climbing will never deliver a would-be simian astronaut onto its lunar surface.

In the previous work, I have identified at least three classical *philosophico-technical* problems that illustrate why computational AI has historically failed, and will continue to fail, to deliver on its "grand challenge" of replicating human mentality in all its raw and electrochemical glory [3] and I will briefly summarize these below.

---

* Strong AI takes seriously the idea that one day machines will be built that can think, be conscious, and have genuine understanding and other cognitive states in virtue of their execution of a particular program; by contrast, weak AI does not aim beyond engineering the mere simulation of (human) intelligent behavior.

12.2.1 Computers Lack (Phenomenal) Consciousness

In science and science fiction, the hope is periodically reignited that a computer system will one day be conscious in virtue of its execution of an appropriate program; in moves toward this goal: World Scientific Publishing produce the *International Journal of Machine Consciousness*. The UK funding body Engineering and Physical Sciences Research Council awarded an Adventure Fund grant of around £500,000 to a team of "roboteers and psychologists" at Essex and Bristol universities, with a goal of instantiating machine consciousness in a "humanoid-like" robot called Cronos through appropriate computational "internal modeling," and already a group of researchers at the University of Reading, led by Kevin Warwick, have claimed that robots they have developed are "as conscious as a slug."

Conversely, in an argument entitled "Dancing with Pixies" (DwP) I demonstrated that if a computer-controlled robot experiences a conscious sensation as it interacts with the world, then an infinitude of "conscious sensation" must be realized in all objects throughout the universe: in this cup of tea that I am drinking as I write, in the seat that I am sitting as I type, and so on. If we reject such "panpsychism," we must reject "machine consciousness."

The underlying thread of the DwP reductio [3–6] derives from positions originally espoused by Hilary Putnam [7], Tim Maudlin [8], and John Searle [9], with subsequent criticism from David Chalmers [10], Colin Klein [11], and Ron Chrisley [12], among others [13].

In the DwP reductio, instead of seeking to secure Putnam's claim that "every open system implements every Finite State Automaton" (FSA) and hence that "psychological states of the brain cannot be functional states of a computer," I establish the weaker result that, over a finite time window, every open physical system implements the particular execution trace of a FSA Q on a specific input vector (I).

That this result leads to panpsychism is clear as, equating FSA $Q(I)$ to a finite computational system that is claimed to instantiate phenomenal states as it executes and employing Putnam's *state-mapping* procedure to map a series of computational states to any arbitrary noncyclic sequence of states, we discover identical computational (and *ex-hypothesis* phenomenal) states lurking in any open physical system (e.g., a rock); then an infinitude of "disembodied experience of conscious sensation" (dancing little pixies) are realized in everything.

Baldly speaking, DwP is a simple *reductio ad absurdum* argument to demonstrate that if the assumed claim is true (that an appropriately programmed computer really does instantiate genuine phenomenal states), then panpsychism is true. However, if, against the backdrop of our immense scientific knowledge of the closed physical world and the corresponding widespread desire to explain everything ultimately in physical terms, we are led to reject panpsychism, then the DwP reductio leads us to reject the claim that any formal computational processes could instantiate phenomenal consciousness.

12.2.2 Computers Lack Genuine Understanding

On June 25, 2012, Google's "Deep Learning" technology was reported by the *New York Times* to have been deployed to categorize unlabeled images. To do this, it was "turned loose on the Internet to learn on its own." Presented with 10 million digital images found in YouTube videos, what did Googles brain do? What millions of humans do with YouTube: looked for cats. Le et al. [14] conjecture, "The focus of this work is to build high-level, class-specific feature detectors from unlabelled images. For instance, we would like to understand if it is possible to build a face detector from only unlabelled images. This approach is inspired by the neuro-scientific conjecture that there exist highly class-specific neurons in the human brain, generally and informally known as 'grandmother neurons.'"

At first sight, if such unsupervised "Deep Learning" algorithms can learn to classify images of "faces," "cats," and "human bodies" from unlabeled images on the Internet, then it would seem that the work must go some way toward demonstrating a genuine form of machine "understanding" (in addition to potentially arbitrating on the age old philosophical question of "natural kinds"*).

However, a thought experiment from the American philosopher John Searle suggests a note of caution. In the now (in)famous *Chinese room argument*, Searle demonstrated how it could be possible to program a computer to *appear* to understand, without the machine *actually* understanding—in his thought experiment Searle famously described how it might be possible to program a computer to communicate perfectly with human interlocutor in a language such as Chinese, without the computer actually understanding anything of the interaction (cf. a small child laughing at a joke he or she does not understand).

---

* In philosophy, the term "natural kind" is used to refer to a "natural" grouping contra an artificial one—an objective contra subjective set. There is considerable debate in analytic philosophy about whether there are any natural kinds at all, because even plausible definitions of very familiar species (such as the cat and the dog) leave the classification of some exemplars ambiguous.

Searle illustrates the point by demonstrating how he could follow the instructions of the program—*in computing parlance, we would say Searle is "dry running" the program*—and carefully manipulating the squiggles and squiggles of the [to him] meaningless Chinese ideographs as instructed by the program, without ever understanding a word of the Chinese ideographic responses the process is so methodically cranking-out.

The essence of the Chinese room argument is that *syntax—the mere mechanical manipulation [as if by computer] of uninterpreted symbols—is* not sufficient for *semantics* (meaning) to emerge. In this way, Searle asserts that no mere computational process can ever bring forth genuine understanding and hence that computation must ultimately fail to fully instantiate mind.*

It is clear that Searle's argument could just as easily target the claim that a Deep Learning network understands the images it so adroitly processes.†

12.2.3 Computers Lack (Mathematical/Creative) Insight

In his book *The Emperor's New Mind,* the Oxford mathematical physicist Sir Roger Penrose deployed Gödel's first incompleteness theorem to argue that, in general, the way mathematicians provide their "unassailable demonstrations" of the truth of certain mathematical assertions is fundamentally non-algorithmic and noncomputational. Gödel's first incompleteness theorem states that "... any effectively generated theory capable of expressing elementary arithmetic cannot be both consistent and complete. In particular, for any consistent, effectively generated formal theory F that proves certain basic arithmetic truths, there is an arithmetical statement that is true, but not provable in the theory." The resulting true but unprovable statement $G(\breve{g})$ is often referred to as "the Gödel sentence" for the theory (albeit there are infinitely many other statements in the theory

---

* See [15] for extended discussion of the Chinese room argument by 20 well-known cognitive scientists and philosophers.
† This philosophical position has recently given an additional empirical weight in a critical follow-up paper from Szegedy et al. [16] in which the researchers demonstrated that "we can cause the network to misclassify an image by applying a certain imperceptible perturbation, which is found by maximizing the networks prediction error. In addition, the specific nature of these perturbations is not a random artefact of learning: the same perturbation can cause a different network, that was trained on a different subset of the dataset, to misclassify the same input"; clearly whatever a Deep Learning network is doing when it has learnt to classify unlabeled data, it has not demonstrated that the "specificity of the 'grandmother neuron' could possibly be learned from unlabeled data" or, expressed more colloquially, that such a network could shed light on the human ability to categorize "cat" images.

that share with the Gödel sentence the property of being true but not provable from the theory).

Arguments based on Gödel's first incompleteness theorem—*initially from John Lucas [17,18], criticized by Paul Benacerraf [19] then subsequently extended, developed and widely popularized by Roger Penrose [20–23]*—typically endeavor to show that for any such formal system F, humans can find the Gödel sentence $G(\breve{g})$, whereas the computation/machine (being itself bound by F) cannot.

In [21], Penrose develops a subtle reformulation of the vanilla argument that purports to show that "the human mathematician can 'see' that the Gödel sentence is true for consistent F even though the consistent F cannot prove $G(\breve{g})$."

NB. A detailed discussion of Penrose's formulation of the Gödelian argument is outside the scope of this chapter—for a critical introduction see [24] and for Penrose's response see [22]—here it is simply important to note that although Gödelian-style arguments purporting to show "computations are not necessary for cognition" have been extensively and vociferously critiqued in the literature (see [25] for a review), interest in them—both positive and negative—still regularly continues to surface (e.g., [26,27]), with Penrose and Hammeroof asserting that recent developments in physics have gone a long way to proving their case [28].

12.3 AI AND ARTIFICIAL STUPIDITY

Taken together, these above arguments undermine the notion that the human mind can be completely instantiated by mere computations; if correct, although computers will undoubtedly get better and better at many particular tasks—say playing chess, driving a car, predicting the weather, and—there will always remain broader aspects of human mentality that future AI systems will not match. Under this conception, there is a "humanity gap" between the human mind and mere "digital computations"; although raw computer power—and concomitant AI software—will continue to improve, the combination of a human mind working alongside a future AI will continue to be more powerful than that future AI system operating on its own; the singularity will never be televised.

Furthermore, it seems to me that without understanding and consciousness of the world and lacking genuine creative (mathematical) insight, any apparently goal-directed behavior in a computer-controlled robot is, at best, merely the reflection of a deep-rooted longing in its designer. Lacking an ability to formulate its own goals, on what basis would a robot set out to subjugate mankind unless, of course, it was explicitly programmed to do so by its (human) engineer? But in that case our underlying apprehension regarding future AI might better reflect the all-too-real concerns surrounding current autonomous weapons systems, than casually reindulging Hollywood's vision of the posthuman "terminator" machine.

Indeed, in my role as one of the AI experts co-opted onto the "International Committee for Robot Arms Control" (ICRAC), I am particularly concerned by the potential military deployment of robotic weapons systems—*systems that can take decisions to militarily engage without human intervention*—precisely because current AI is still very lacking and because of the underlying potential of poorly designed interacting autonomous systems to rapidly escalate situations to catastrophic conclusions; in my view such systems all too easily exhibit genuine "artificial stupidity."

I am particularly skeptical that current and foreseeable AI technology can enable autonomous weapons systems to *reliably* comply with extant obligations under the International Humanitarian Law, specifically three core obligations: (1) to identify combatants from noncombatants, (2) to make nuanced decisions regarding proportionate responses to a complex military situation, and (3) to arbitrate on military or moral necessity regarding when to apply force.

The extreme difficulty in lawfully identifying combatants from noncombatants is powerfully highlighted in the following example from the Human Rights Watch report *Losing Humanity: The Case against Killer Robots* [29]:

> … According to philosopher Marcello Guarini and computer scientist Paul Bello, "[i]n a context where we cannot assume that everyone present is a combatant, then we have to figure out who is a combatant and who is not. This frequently requires the attribution of intention." One way to determine intention is to understand an individuals emotional state, something that can only be done if the soldier has emotions. Guarini and Bello continue, "A system without emotion could not predict the emotions or action of others based on its own states because it has no emotional states." Roboticist Noel Sharkey echoes this argument: "Humans understand one another in a way that machines cannot. Cues can

be very subtle, and there are an infinite number of circumstances where lethal force is inappropriate."

For example, a frightened mother may run after her two children and yell at them to stop playing with toy guns near a soldier. A human soldier could identify with the mothers fear and the childrens game and thus recognize their intentions as harmless, while a fully autonomous weapon might see only a person running toward it and two armed individuals. The former would hold fire, and the latter might launch an attack. Technological fixes could not give fully autonomous weapons the ability to relate to and understand humans that is needed to pick up on such cues.

In addition to the technical challenges of meeting obligations under the International Humanitarian Law, whenever autonomous systems interact without human supervision, there is also a very real danger of catastrophic unintended escalation as underlying problems of "artificial stupidity" forcefully come to bear.

A light-hearted example demonstrating just how easily autonomous systems can rapidly escalate situations out of control occurred in April 2011, when Peter Lawrence's book *The Making of a Fly* was auto-priced upward by two "trader-bots" competing against each other in the Amazon reseller marketplace. The result of this process is that Lawrence can now comfortably boast that his modest scholarly tract—first published in 1992 and currently out of print—was once valued by one of the biggest and most respected companies on the Earth at $23,698,655.93 (plus $3.99 shipping).

As stark contrast, in Machine gun-toting robots deployed on DMZ, a report in *Stars and Stripes* magazine (July 12, 2010), Jon Rabiroff outlines the following terrifying scenario:

> DEMILITARIZED ZONE, Korea Security along the DMZ has gone high-tech, as South Korea has quietly installed a number of machine gun-armed robots to serve as the first line of defense against the potential advance of North Korean soldiers.
>
> The stationary robots which look like a cross between a traffic signal and a tourist-trap telescope are more drone than Terminator in concept, operated remotely just outside the southern boundary of the DMZ by humans in a nearby command center.
>
> Officials refuse to say how many or where the robots have been deployed along the heavily fortified border between the two

Koreas, but did say they were installed late last month and will be operated on an experimental basis through the end of the year.

South Korean military officials will then decide how many, if any, robots they want complementing the soldiers who man the area adjacent to the 2.5 mile-wide DMZ, which stretches 160 miles across the peninsula.

"The robots are not being deployed to replace or free up human soldiers," said Huh Kwang-hak, a spokesman for Samsung Techwin, the manufacturer of the SGR-1 robot. "Rather, they will become part of the defense team with our human soldiers. Human soldiers can easily fall asleep or allow for the depreciation of their concentration over time," he said. "But these robots have automatic surveillance, which doesnt leave room for anything resembling human laziness. They also wont have any fear (of) enemy attackers on the front lines."

South Korea Ministry of National Defense spokesman Kwon Ki-hyeon said his agency is overseeing the project so he could not comment on the DMZ robot experiment. He referred questions to Samsung Techwin.

Huh said no government officials would talk about the robots: "This experimental project is highly classified."

With armed robot border guards patrolling one side of the DMZ and the potential for North Korea to respond in kind, the darkly dystopian "science fiction" vision of two quasi-autonomous robot armies squaring up to each other begins to look all too possible; furthermore, given that one of the protagonists is an unstable nuclear armed state, the unintended dangers from a relatively minor military transgression, say a minor border incursion, escalating into a very serious, potentially nuclear, confrontation, begin to look alarmingly possible.

12.4 BODY IN QUESTION

I believe that the DwP reductio, John Searle's "Chinese room argument" and Roger Penrose' reflections on the noncomputable nature of mathematical insight suggest that we need to move away from purely computational explanations of cognitive processes and instead reflect on how meaning, teleology, and human creative processes are fundamentally grounded in the human body, society, and the world, obliging us, in turn, to take issues of embodiment—the body and our social embedding—much more seriously. And such a *strong* notion of embodiment most certainly cannot be realized by simply co-opting a putative computational creative system into a conventional *tin can robot.**

As Slawomir Nasuto and I set out in our recent discussion of *biologically controlled animals*[†] and the so-called *zombie animals*[‡] (two examples carefully chosen to lie at polar ends of the spectrum of possible engineered robotic/cyborg systems), because the induced behavioral couplings therein are not the effect of the intrinsic "nervous" system's constraints (metabolic or otherwise) at any level, *a fortiori*, merely instantiating appropriate sensorimotor coupling is not sufficient to instantiate any meaningful intentional states [31].

On the contrary, in both *zombie animals* and *biologically controlled animats*, the sensorimotor couplings are actually the *cause* of extrinsic metabolic demands (made via the experimenter's externally directed manipulations). But because the experimenter drives the sensorimotor couplings in a completely arbitrary way (from the perspective of the intrinsic metabolic needs of animal or its cellular constituents), the actual causal relationship between the bodily milieu and the motor actions and sensory readings can never be genuinely and appropriately coupled. Hence, our conclusion (ibid) that *only the right type and directionality of sensorimotor couplings* can ultimately lead to genuine understanding and intentionality.

In light of such concerns, and until the challenges of the CRA, DwP and the mystery of mathematical insight have been *fully* met and the role of embodiment more strongly engaged (such that the neurons, brain, and body fully interact with other bodies, world, and society), I suggest a note of caution in labeling any artificial system as "strongly intelligent"—*a computational mind*—in its own right; any "cognition" displayed therein being merely a projection of its engineer's intellect, esthetic judgment, and desire.

12.5 CONCLUSION

Without having to fantasize that it has now (or will ever) reached the level of superhuman intelligence that Professors Warwick and Hawking have graphically warned us of, the all-too-real-world example of armed robots

---

* Whereby an appropriate AI is simply bolted onto a classical robot body and the particular material of that "embodiment" is effectively unimportant.

[†] Robots controlled by a cultured array of real biological neurons.

[‡] For example, an animal whose behavior is "remotely controlled" by an external experimenter, say by optogenetics; see also Gradinaru et al. [30], who used optogenetic techniques to stimulate neurons selectively, inducing motor behavior without requiring conditioning.

(as described earlier) precisely illustrate why it is easy to concur that already current AI systems pose a real "existential threat" to humanity—the threat of *artificial stupidity*. For this reason, in May 2014, members of the ICRAC gathered in Geneva to participate in the first multilateral meeting ever held on lethal autonomous weapons systems (LAWS)—a debate that continues to this day at the very highest levels of the United Nations, in a firm, but refracted, echo of Warwick and Hawking on AI—I think we should be very concerned.

Nonetheless, it is equally obvious that even current-state AI has a rich potential to transform society from the "trivial" replacement of tedious human labor (e.g., by controlling robots to clean the floor and mow the lawn) to a more complex new role as an international social facilitator (by helping people communicate more easily by instantaneously offering an approximate translation from one language to another), to helping the State make substantially better use of scarce public resources (e.g., one project that I was personally closely involved with—the UKPLC "SpendInsight" system—has been recently evaluated by the UK National Audit Office and used to identify potential annual saving in the UK National Health Service purchasing budget in excess of £500 million per annum [32,33]; clearly if such savings were realized, they would buy a significant number of additional frontline doctors, nurses, and drugs).

Already, post "Lighthill," post "connectionist winter," and post "Terminatorblues," the recent practical realization of ambitious *real-world, nouveau AI, machine learning, big data systems* is tempting the engineered geek in me with too many lucrative, new, and seductive headline images, any one of which could so easily prompt me to fall headstrong-in-love with AI again.

REFERENCES

1. Kurzweil, R. (2005). *The Singularity Is Near: When Humans Transcend Biology*. Viking, London.
2. Dreyfus, H.L. (2012). A history of first step fallacies. *Minds and Machines* 22(2—special issue "Philosophy of AI" Müller, V.C., ed.): 87–99.
3. Bishop, J.M. (2009). A cognitive computation fallacy? Cognition, computations and panpsychism. *Cognitive Computation* 1(3): 221–233.
4. Bishop, J.M. (2002). Dancing with pixies: Strong artificial intelligence and panpsychism. In: Preston, J., & Bishop, J.M. (eds.), *Views into the Chinese Room: New Essays on Searle and Artificial Intelligence*. Oxford University Press, Oxford.
5. Bishop, J.M. (2005). Can computers feel? *The AISB Quarterly* 199: 6.

6. Bishop, J.M. (2009). Why computers can't feel pain. *Minds and Machines* 19(4): 507–516.
7. Putnam, H. (1988). *Representation and Reality*. Bradford Books, Cambridge, MA.
8. Maudlin, T. (1989). Computation and consciousness. *Journal of Philosophy* 86: 407–432.
9. Searle, J. (1990). Is the brain a digital computer? *Proceedings of the American Philosophical Association* 64: 21–37.
10. Chalmers, D.J. (1996). *The Conscious Mind: In Search of a Fundamental Theory*. Oxford University Press, Oxford.
11. Klein, C. (2004). *Maudlin on Computation* (working paper).
12. Chrisley, R. (2006, April 4–8). Counterfactual computational vehicles of consciousness. In: *Toward a Science of Consciousness*. Tucson Convention Center, Tucson, AZ.
13. Minds and Machines. (1994). Special issue: What is computation? *Minds and Machines* 4(4).
14. Le, Q., Ranzato, M.A., Monga, R., Devin, M., Chen, K., Corrado, G., Dean, J., & Ng. A. (2012). Building high-level features using large scale unsupervised learning. In: *Proceedings of the 29th International Conference in Machine Learning*. Scotland.
15. Preston, J., & Bishop, J.M. (eds.). (2002). *Views into the Chinese Room: New Essays on Searle and Artificial Intelligence*. Oxford University Press, Oxford.
16. Szegedy, C., Zaremba, W., Sutskever, I., Bruna, J., Erhan, D., Goodfellow, I.J., & Fergus, R. (2013). Intriguing properties of neural networks. In: *Proceedings of the Inc. Conf. International Conference on Learning Representations*.
17. Lucas, J.R. (1962). Minds, machines & Godel. *Philosophy* 36: 112–127.
18. Lucas, J.R. (1968). Satan stultified: A rejoinder to Paul Benacerraf. *Monist* 52: 145–158.
19. Benacerraf, P. (1967). God, the devil & Godel. *Monist* 51: 9–32.
20. Penrose, R. (1989). *The Emperor's New Mind: Concerning Computers, Minds, and the Laws of Physics*. Oxford University Press, Oxford.
21. Penrose, R. (1994). *Shadows of the Mind: A Search for the Missing Science of Consciousness*. Oxford University Press, Oxford.
22. Penrose, R. (1996). Beyond the doubting of a shadow: A reply to commentaries on "Shadows of the Mind." *PSYCHE* 2(23).
23. Penrose, R. (1997). On understanding understanding. *International Studies in the Philosophy of Science* 11(1): 7–20.
24. Chalmers, D.J. (1995). Minds, machines and mathematics: A review of "shadows of the mind" by Roger Penrose. *PSYCHE* 2(9).
25. PSYCHE. (1995). Symposium on Roger Penrose's shadows of the mind. *PSYCHE* 2. http://psyche.cs.monash.edu.au/psyche-index-v2.html.
26. Bringsjord, S., & Xiao, H. (2000). A refutation of Penrose's Gödelian case against artificial intelligence. *Journal of Experimental & Theoretical Artificial Intelligence* 12: 307–329.

27. Tassinari, R.P., & D'Ottaviano, I.M.L. (2007). Cogito ergo sum non machina! About Gödel's first incompleteness theorem and Turing machines. *CLE e-Prints* 7(3).

28. Penrose, R., & Hammeroff, S. (2014). Consciousness in the universe: A review of the Orch OR theory. *Physics of Life Reviews* 11(1): 39–78.

29. Human Rights Watch. (2012, November 19). *Losing Humanity: The Case against Killer Robots.* Human Rights Watch Report.

30. Gradinaru, V., Thompson, K.R., Zhang, F., Mogri, M., Kay, K., Schneider, M.B., & Deisseroth, K. (2007). Targeting and readout strategies for fast optical neural control in vitro and in vivo. *Journal of Neuroscience* 26, 27(52): 14231–14238.

31. Nasuto, S.J., & Bishop, J.M., (2013). Of (zombie) mice and animats. In: Müller, V.C. (ed.), *Philosophy and Theory of Artificial Intelligence. Studies in Applied Philosophy, Epistemology and Rational Ethics* 5: 85–106. Springer, Berlin, Germany.

32. National Audit Office. (2011). *The Procurement of Consumables by NHS Hospital Trusts* [online]. http://www.nao.org.uk/publications/1011/nhs_procurement.aspx (accessed July 16, 2012).

33. Roberts, P.J., Mitchell, R.J., Ruiz, V.F., & Bishop, J.M. (2014). Classification in e-procurement. *International Journal of Applied Pattern Recognition* 1(3): 298–314.

34. Mnih, V., Kavukcuoglu, K., Silver, D., Rusu, A.A., Veness, J., Bellemare. M.G., Graves, A. et al. (2015). Human-level control through deep reinforcement learning. *Nature* 518, 529–533.

35. Searle, J. (1980). Minds, brains and programs. *Behavioral and Brain Sciences* 3(3): 417–457.

Index

Note: Locator followed by '*f*' and '*t*' denotes figure and table in the text

Milton Keynes UK
Ingram Content Group UK Ltd.
UKHW040447071024
449327UK00020B/1061